8/12

635

CrL

Please renew/return this item by the last date shown.

So that your telephone call is charged at local rate,
please call the numbers as set out below:

	From Area codes 01923 or 020:	From the rest of Herts:
Renewals:	01923 471373	01438 737373
Enquiries:	01923 471333	01438 737333
Textphone:	01923 471599	01438 737599

L32   www.hertsdirect.org/librarycatalogue

50 Years of Garden Machinery

50 Years of Garden Machinery

Brian Bell MBE

FARMING PRESS

ISBN 0 85236 301 X

A catalogue record for this book is available from the British Library

Published by Farming Press Books
Wharfedale Road, Ipswich IP1 4LG, United Kingdom

Distributed in North America
by Diamond Farm Enterprises,
Box 537, Alexandria Bay, NY 13607, USA

Cover photographs
Front Landmaster garden cultivator
Back Allen motor scythe and an 18 in. Hayter Hayterette

Frontispiece
An 18 in. Hayter mower

Cover design by Andrew Thistlethwaite
Typeset by Galleon Photosetting, Ipswich
Printed and bound in Great Britain by Butler & Tanner Ltd, Frome and London

Contents

There is a colour section between pages 120 and 121

Acknowledgements

I am indebted to many people and companies for their help in compiling this book. Gordon Addison, Denis Allman, Fred Chapman, Ivan Clarke, Brian Hurtley, Mr and Mrs J. Knifton, Kim McFie, Joe Paget, Bob Rendle and Roger Tombs have provided me with a wealth of information and photographs. Additional photographs have been loaned by L.G. Bassett, John Briscoe, Delia Chinnery, Stuart Gibbard, Brian Harraway, Lynn Hilton, Anthony Marshall, Frank Moore, Joe Pendal, Helen Pocock, Wally Smith and Lorraine Surtees.

Conversion Table

It is usual to provide conversion tables for those less familiar with the metric system. In keeping with the nostalgic flavour of this book, the information below is offered to help those readers who may be too young to remember the imperial days of pounds, shillings and pence.

1 gallon	= 4.6 litres
1 inch	= 25.4 mm
1 foot	= 300 mm
1 yard	= 910 mm
1 acre	= 0.4 hectare
1 cwt	= 50.8 kg
1 ton	= 1,016 kg
1 shilling	= 5 pence
1 pound	= 20 shillings

Introduction

The mechanisation of gardening was already underway in the mid 1930s when a wooden wheelbarrow cost 18s 6d and a 23 in. cut side-wheel mower could be purchased for £2 12s 3d (£2.61). In the last fifty years the development of garden machinery has brought an end to back-ache and tired legs after a hard day in the garden of heavy digging with a spade or pushing a hand mower.

This book traces the changes in powered equipment since 1945 for the many and varied gardening tasks, including tilling the soil, mowing lawns and trimming the garden hedge. By the 1940s, petrol engines and electric motors had replaced horse, donkey and steam power on the lawns around our stately homes, and hand mowers had made life easier for the domestic gardener. Few hand mowers are used in the 1990s when, according to need, we can buy a self-propelled or a ride-on rotary mower with an electric starter. The spade and fork still have a place in the garden but even here various shapes and sizes of garden tractor have taken the drudgery from digging, planting and hoeing the vegetable patch.

Garden machinery has been made by numerous companies during the last fifty years, and brief histories of Atco, Hayter, Qualcast, Ransomes, Suffolk Iron Foundry, Trusty and other prominent manufacturers are recorded on the following pages. A book of this size can allow only a snapshot of the many hundreds of garden machines made since the mid 1940s. Some were sold in tens of thousands, and others were less successful. A few of them would certainly fail to gain official approval in the safety-conscious 1990s.

BRIAN BELL

1 Garden Tractors

For centuries, gardeners and smallholders relied on horse-drawn implements and hand tools to prepare seedbeds, kill weeds and harvest their crops but by the mid 1940s self-propelled pedestrian-controlled garden tractors were in common use. Hand-propelled implements such as the one- or two-wheeled Planet Junior push hoe from America were used by vegetable growers to sow, hoe and cultivate their crops. Many market gardeners with large areas under cultivation were the proud owners of a pedestrian-controlled or walking tractor with a petrol engine between 1 and 6 hp.

British-made two-wheel pedestrian-controlled garden tractors including the Auto-Culto, B.M.B, British Anzani, Garner, Howard, Rototiller and Trusty were a common sight on smallholdings and in large gardens by the late 1940s. A few American-built garden tractors including the Planet Junior and Gravely were also available in Great Britain at this time.

There were two distinct groups of garden or walking tractors. The Howard Gem and Landmaster were classed as self-propelled rotary cultivators. Others ranging from the $1\frac{1}{2}$ hp Farmers' Boy to the 6 hp British Anzani Iron

Plate 1.1   A single-furrow plough was one of the attachments made for the Trusty Earthquake rotary cultivator.

Horse were used for ploughing, cultivating, row crop work, and many of them could pull a small trailer.

Pedestrian-controlled garden tractors gradually lost favour in the early 1950s as more and more market gardeners and smallholders traded up to a four-wheeled ride-on tractor such as the Garner, Gunsmith and Trusty Steed. Today two-wheel garden cultivators, now far more efficient than their predecessors, remain popular with gardeners growing large quantities of fruit and vegetables.

Allen & Simmonds, who used 'Pistons, Reading' as their telegraphic address, made the first Auto-Culto garden tractors at Reading in 1926. By 1940 there was a choice of Villiers $1\frac{1}{2}$, $2\frac{1}{2}$ or $3\frac{1}{2}$ hp two-stroke engines to propel the steel-wheeled Auto-Culto, which had a dog clutch to engage the gear drive arrangement. At this time Allen & Simmonds also made the Auto-Culto Junior. Used to cultivate between rows of plants and small trees, it was only 12 in. wide and cost £45. The 1940 Auto-Culto catalogue apologised that there might be some delay in delivery and requested customers to bear in mind that the company was busy executing urgent orders for the Admiralty and Ministry of Supply.

An unusual starting device appeared on Auto-Culto tractors in 1949. The principle was similar to a kick start with a push handle and strong return spring attached to the engine cowling. The handle, which could well have been used for a garden fork, was pushed down to turn the

DE LUXE MODEL. 4 h.p. 4-stroke engine, petrol or paraffin, power turning, two forward speeds and reverse, centrifugal friction clutch

The Auto Culto range is fully comprehensive. The De Luxe and "M" Models will operate all Auto Culto attachments. A range of attachments is also available for the "Midget." Let us give you a demonstration free and without obligation. H.P. terms available. For details write to "The Pioneers of the Two Wheel Tractor."

ALLEN & SIMMONDS (AUTO CULTO) LTD. READING

Fig 1.1   The Auto-Culto garden tractor range in the early 1950s.

Plate 1.2   This $2\frac{1}{2}$ hp Auto-Culto cost £75 12s 0d in 1940. The price included a set of spanners, grease gun and tool box.

crankshaft and the makers claimed that with the help of the spring it was not difficult to overcome cylinder compression and start the engine.

Allen & Simmonds marked the 1951 Festival of Britain with the Festival Auto-Culto, designed for fruit and market garden work. The Festival, which cost about £140, had a $4\frac{1}{2}$ hp Villiers four-stroke petrol or paraffin engine, a two forward and reverse gearbox and dog clutches in both wheels for power turns. Implements for the Festival, which could be used on the Auto-Culto Midget, Model L and Model M, included a plough, rotary cultivator, sprayer and grass mower.

At the 1954 Smithfield Show Allen & Simmonds launched the Autogardener, which cost

£75 and had a selection of implements including a plough, rotary cultivator and cutter bar. The specification included a four-stroke Villiers engine which gave six hours' working on a gallon of fuel, a lever-operated plate clutch and a three-speed gearbox with a top speed of $1\frac{1}{2}$ mph. The cultivating rotor was driven by a three-speed power take-off shaft running at 120, 250 or 350 rpm. The power shaft could also be used to drive an air compressor used for paint spraying and the application of weed killers, pesticides, etc.

An improved Universal Autogardener was made in 1962 with an extended implement range including a potato lifter, log saw, pump, hedge cutter and an inter-row hoe on a front or rear tool frame.

The 1.2 hp four-stroke engined Horti-Culto rotary cultivator advertised in 1959 as the complete home gardener cost £57. It was also used for hoeing, ploughing, clearing scrub, mowing lawns and trimming hedges. The Auto-Culto 600, which cost £175, was another Allen & Simmonds product in the late 1950s.

Unlike many of its contemporaries, Auto-Cultos were still made in the mid 1970s by Jalo Engineering at Wimborne, who had taken over the business from Allen & Simmonds. Four models were listed in 1976: the Auto-Culto 55, Auto-Culto Mark IX, the Autogardener and the Midgi Culto.

The rotary cultivator at the front of the Auto-Culto 55 was driven by a $3\frac{1}{2}$ hp Norton Villiers four-stroke engine. Power was transmitted by a vee-belt with a tensioner pulley as a clutch and an enclosed roller chain to the standard 12 in. rotor which could be extended to 48 in.

The other models had pneumatic tyres and a rotary cultivator unit under the handlebars, which was interchangeable with a tool bar, plough, cutter bar, etc. The 6 hp Villiers/J.A.P-powered Mark IX had a dry plate clutch, three forward gears and reverse and two power take-off shafts. The Autogardener with a 3 hp Villiers Mark F 15 four-stroke engine and a three-speed power take-off had a top speed of 2 mph. The 150 cc Villiers-engined Midgi Culto had a single plate clutch and three-speed gearbox to drive the wheels and 14 in. rotor.

The British Anzani Engineering Co. at Hampton Hill in Middlesex, manufacturers of aeroplane and speed boat engines, introduced the British Anzani Iron Horse in 1940. It had a 6 hp J.A.P four-stroke air-cooled engine, oil bath air cleaner and Wico magneto. A centrifugal clutch engaged the drive to the steel wheels through a three forward and reverse gearbox when the engine reached a speed of 750 rpm.

A centrifugal clutch has a pair of shoes with friction linings attached to a disc on the engine crankshaft. A spring-loaded linkage allows the shoes to move outwards as the crankshaft gains speed and drive is engaged when they come into contact with the inner face of a drum on the output shaft. When engine speed falls below a pre-set level, the springs withdraw the shoes from the drum and drive is disconnected.

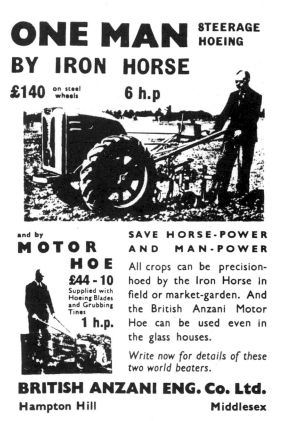

Fig 1.2 British Anzani garden tractor.

Hand-lever-operated independent dog clutches in the drive to the wheels enabled the Iron Horse to make very sharp turns on the headland. Caution was required as the engine power would swing the handlebars round with some force and could easily knock the operator off his feet. Spade lug wheels were standard equipment and track width was adjustable from 24 to 36 in. for ploughing and row-crop work. Optional extras included a side-mounted belt pulley suitable for a saw bench and other stationary equipment. The British Anzani Iron Horse cost £107 16s 0d in 1940, and pneumatic tyres were an extra £10; the price had increased to £130 by 1945.

Hire purchase terms for the British Anzani in 1948 required a £35 deposit with two years to pay the balance. Alternatively the Iron Horse and a selection of implements could be hired for £3 10s 0d per week.

Plate 1.3 The 6 hp British Anzani Iron Horse was introduced in 1940.

British Motor Boats, a London-based company, was importing small marine engines and Simplicity two-wheeled garden tractors from America in the 1930s. The tractors were sold in Great Britain as the B.M.B Plow-Mate, Cult-Mate and Hoe-Mate. British Motor Boats moved to Banbury at the outbreak of war and soon after this a shortage of supplies from America resulted in limited production of the Hoe-Mate by Shillans Engineering at Banbury.

Brockhouse Engineering bought British Motor Boats in the mid 1940s, and production was transferred to Crossens in Lancashire. Improved versions of the Plow-Mate, Cult-Mate and Hoe-Mate introduced in 1947 were made until B.M.B walking tractor production ceased around 1956.

The Plow-Mate, the largest of the three B.M.B walking tractors, had a 6 hp J.A.P air-cooled four-stroke engine, a gearbox with two forward gears and reverse, a power take-off and independent brakes. A 6 hp Briggs & Stratton model ZZ four-stroke engine was fitted to later models of the Plow-Mate. The flat belt drive from the

engine to gearbox was engaged with an idler pulley operated by an over-centre lever on the handlebars. Spade lug wheels were standard with pneumatic tyres an optional extra and the wheel track could be adjusted for row-crop work.

The first British-built Cult-Mate tractors made by Shillans Engineering had a 3 hp engine but this was soon replaced with a B.S.A. $3\frac{1}{2}$ hp power unit. Forward speed was between $\frac{1}{2}$ and $2\frac{1}{2}$ mph and the flat belt drive to the gearbox was engaged with the same mechanism as that used on the Plow-Mate. Over-run ratchets in both wheel hubs provided differential action when turning on headlands, etc. A toolbar with two castor wheels for inter-row work had a separate steering lever between the handlebars to help the operator to work close to rows of plants without damaging them. Sales literature suggested the Cult-Mate was so easy to handle that a child could manage it and that no operator could fail to appreciate its manoeuvrability or the speed and ease with which the many row-crop operations could be performed without back-breaking daily fatigue.

The Hoe-Mate with its air-cooled 1 hp two-stroke Spryt engine, later replaced with a $1\frac{1}{2}$ hp B.M.B power unit, was the smallest B.M.B garden tractor. A lever on the handlebars tensioned the vee-belt drive from the engine, and differential action was provided by over-run ratchets in

Plate 1.4 The B.M.B Cult-Mate had a top speed of $2\frac{1}{2}$ mph.

Plate 1.5   The Hoe-mate was the baby of the B.M.B range.

the wheel hubs. The Hoe-Mate was used for cultivating, hoeing, discing, ridging, etc. and a 30 in. cutter bar suitable for cutting hay or long grass added to the versatility of this little tractor.

A sales leaflet claimed the Hoe-Mate would work all day on less than a gallon of petrol and a testimonial letter from one owner stated that he bought one because a 'girl with a Hoe-Mate could do as much work as three or four men with hand hoes'.

Landmaster rotary cultivators were made by Byron Horticultural Engineering at Hucknall near Nottingham in the early 1950s. The company name had changed to Landmaster Ltd by 1954, when advertising material advised potential customers that although Landmaster walking tractors were designed for market gardeners, they could also take the drudgery from cultivating the domestic vegetable patch.

By the mid 1950s the blue and white Landmaster rotary hoe was a serious competitor for the Howard Rotavator. The smallest model, the Gardenmaster 80 with a 1 hp two-stroke J.A.P engine and attachments that included a rotary cultivator, spinweeder, hoe blades and grass cutter, was popular with vegetable gardeners. A flexible drive cable facilitated the

9 REASONS WHY THE B.M.B. LIGHT TRACTOR *is your best investment*

1. Power-Turning.
2. Unit construction gear box and reduction box.
3. Clutch drive of exclusive B.M.B. design.
4. Simplicity of control and operation.
5. Complete range of quickly interchangeable implements.
6. Low in initial cost and maintenance.
7. Power Take-off.
8. Interchangeable rubber-tyred and steel lug wheels.
9. Adjustable wheel centres.

6 h.p. "PLOWMATE"   3 h.p. "CULTMATE"
        1½ h.p. "HOEMATE"

Fig 1.3   The B.M.B Plow-Mate.

Fig 1.4   *The Landmaster rotary hoe was exhibited at the 1953 Smithfield Show.*

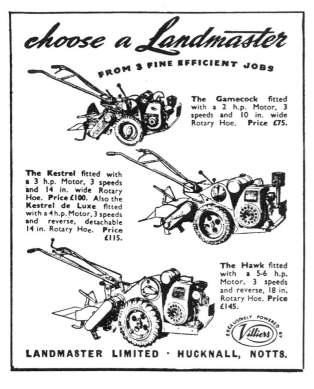

Fig 1.5   The 1954 range of Landmaster garden tractors.

use of a small hedge trimmer and a drill chuck for the handyman to drill holes in wood and metal. The Gardenmaster 80 had acquired a $1\frac{1}{2}$ hp engine by 1959, when it cost £49 complete with rotor head and tool kit.

The Landmaster L150 rotary cultivator, priced at £95, was exhibited for the first time at the 1959 Chelsea Flower Show. Power from a J.A.P engine was transmitted by a two-speed vee-belt drive, reduction gears and roller chain to the rotor shaft. Fuel consumption was between 1 and 3 pints per hour. The cultivating rotor could be extended to give working widths from $7\frac{1}{2}$ to 27 in., and after the rotor was replaced with a pair of driving wheels it could be used with many attachments including a toolbar, seeder unit, centrifugal pump and reversible plough. The L 150 was also sold as a garden tractor with a single furrow plough for £119 16s 0d.

There were five models of the Gardenmaster Power Gardener for garden and allotment work in 1963. The $\frac{3}{4}$ hp two-stroke Gardenmaster 34 with 12 in. wide digging blades and spinweeder cost £41 10s 0d. Next in line were the $1\frac{1}{2}$ hp two-stroke Gardenmaster 80 at £49 15s 0d, the 3 hp

Fig 1.6   This $\frac{3}{4}$ hp Gardenmaster Power Gardener 34, considered ideal for up to $\frac{1}{3}$ acre, cost £37 10s 0d in 1959.

four-stroke 85 and the sturdier 3 hp Gardenmaster 95, which cost £62 10s 0d. All were supplied with a spinweeder and a 12 in. rotor with additional blades to give a maximum digging width of 18 in. The Gardenmaster 100, the largest Power Gardener, cost £69 10s 0d. It had a 3 hp four-stroke engine with a lever-operated clutch to disengage drive to the 18 in. wide digging blades. Attachments for Landmaster Power Gardeners included a grass mower, water pump, Allman sprayer and flexible drive for Tarpen or Heli-Strand Tools hedge trimmers.

Other Landmaster garden tractors in the mid 1960s included an improved 150, the 243 and the Mo'dig. A sales leaflet suggested that the $2\frac{1}{2}$ hp two-stroke Mo'dig, similar in appearance to the Gardenmaster 80, was a new dimension in power gardening. As its name suggests, the

Plate 1.6 The 4 hp Landmaster 243, which cost £84 10s 0d in 1964, was suitable for inter-row hoeing and deep cultivations.

Mo'dig could be used as a rotary mower or a power cultivator, and complete with mower, grass collecting bag, rotary lawn rake, spin-weeder and cultivator head, it cost £69.

The 1964 Landmaster 150 with a 4 hp two- or four-stroke Clinton engine and a 33 in. rotary cultivator cost £123 10s 0d. After the rotor was replaced with a pair of pneumatic wheels, the 150 could be used for ploughing, ridging, grass cutting and trailer work.

The Landmaster 243 with a 24 in. cultivating rotor had a 4 hp four-stroke engine, two forward gears and reverse and adjustable handlebars. Sales literature for the 243, which cost £84 10s 0d, described it as two machines in one because the engine could be moved forward over the cultivating rotor for added penetration or backwards over the wheels when used for inter-row hoeing, ridging, etc. The 3 hp L120 rotary cultivator was another late 1960s Landmaster machine with the usual range of implements and accessories.

Landmaster moved to Poole in 1971 and became part of the Wolseley Webb group in 1980. Wolseley Webb were acquired by Qualcast in 1984. The 1981 Landmaster range consisted of the 4 hp Landmaster 140 (similar in appearance to the Merry Tiller), also the 3 hp Landmaster Lion Cub and the 3 hp L88. The Lion Cub with one pair of slasher rotors and the L88 with two pairs followed the design of earlier Landmaster garden tractors.

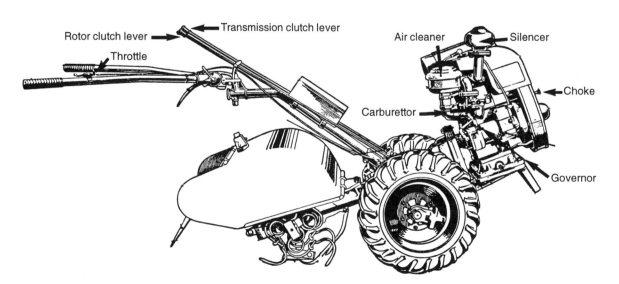

Fig 1.7   The Clifford Model A Mark I Rotary cultivator.

Plate 1.7   The price in 1950 of the Clifford Model B rotary cultivator and inter-row hoe with pneumatic tyres and 12 in. cultivating rotor was £126 10s 0d.

Precision engineers Clifford Aero and Auto Ltd were established in 1912. This company, later renamed Clifford Cultivators, started making garden cultivators in 1947. There were three versions of Clifford Model A rotary cultivator. The Model B was designed for greenhouse and row-crop work. Implements included a plough, mower, sprayer, ridger, hoes, potato spinner and trailer.

The Model A Mark I had a 5 hp air-cooled J.A.P engine with dog clutches to engage the single speed worm and wheel drive to the land wheels and direct drive to the spring tine cultivating rotor. The one gallon petrol tank held enough fuel for about 2½ hours' work. The Mark I with a 16 in. wide rotor cost £100 and a 22 in. rotor added £5 to the price.

The Mark II had the same 5 hp J.A.P engine but it was more sophisticated with a friction clutch operated by a lever on the handlebars and a gearbox provided forward speeds of 1 and 2 mph. The standard machine with a 22 in. working width cost £137 10s 0d and there was a reduction of £1 15s 0d for the Mark II with a 16 in. rotor. Reverse gear and wheel clutches for power turning were optional extras.

The Mark III was the most expensive Clifford Model A rotary cultivator. Standard equipment included a hand-operated friction clutch, a two forward and reverse gearbox and independent wheel clutches for easy turning on headlands. A gear box lever was used to select forward speed, and reverse was engaged with a separate lever on the handlebars. Larger diameter wheels (plate 3.26), which increased the forward speed by 50%, were an optional fitting for Model A rotary cultivators. The 24 in. Mark III Clifford rotary cultivator cost £160.

The 5 hp Clifford Model B was a combined rotary cultivator and inter-row hoe with a friction clutch, two forward speeds and a 12 in. rotor. A gallon of petrol was said to be sufficient for about three hours' work.

A second generation of rotary cultivators was made by Clifford Cultivators at West Horndon in the mid 1950s. The Mark I with a Villiers two-stroke power unit was a compact gear-driven machine that could dig, hoe, ridge and cut grass and trim hedges. Similar to the earlier models, the 7½ hp Mark IV Clifford rotary cultivator introduced at the 1953 Smithfield Show had a worm drive to the 16 or 22 in. rotor with

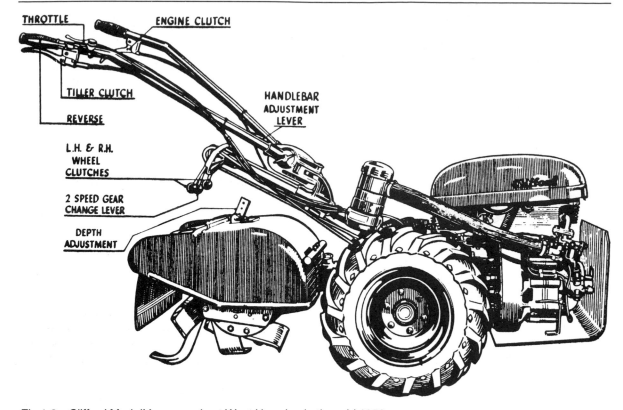

THROTTLE

ENGINE CLUTCH

TILLER CLUTCH

HANDLEBAR
ADJUSTMENT
LEVER

REVERSE

L.H. & R.H.
WHEEL
CLUTCHES

2 SPEED GEAR
CHANGE LEVER

DEPTH
ADJUSTMENT

Fig 1.8   Clifford Mark IV was made at West Horndon in the mid 1950s.

Plate 1.8   The Mark 1 Clifford Cultivator was made at West Horndon and cost £101 10s 0d in 1958.

TRUSTY
- keeps on working

THE BRITISH LIGHT TRACTOR WITH A NAME FOR

Reliability

This dependable power unit incorporates all those refinements of design which make for super-efficiency, easy running and light handling. The sturdy tireless Trusty tackles every job it is put to, from ploughing to hauling — and is always fit for more. Look into its capabilities. You'll find the 'TRUSTY' and its tools invaluable.

EFFICIENT
VERSATILE
DEPENDABLE

Trusty leads the way with
● Ample reserve of the right type of power ● Patent Automatic Clutch ● Patent Swinging Draw-bar ● 'Silky' start, smooth running and self-steering ● Scientific weight distribution ● Torque-reaction for soil penetration ● Adjustable width ● Specially designed tool for *every* job ● Implements speedily, easily changed.

Price from £125 to £140
Implements extra
IMMEDIATE DELIVERY

Fig 1.9   Trusty garden tractor advertisement.

rigid blades or spring tines. There was a choice of power units, either a J.A.P 600 cc or B.S.A. 500 cc, both with dry sumps. Lubrication was by means of an oil tank beneath the engine hood with a pump to force feed oil to the moving parts.

A diesel-engined version of the Clifford Mark IV was launched at the 1959 Chelsea Flower Show. By December of that year Clifford Cultivators had amalgamated with Howard Rotavators and their machines carried a Howard-Clifford logo.

Mayfield Engineering at Dorking made three models of garden tractor during the 1950s. They all had four-stroke engines and three-speed gearboxes with an optional reverse gear. An advertisement informed potential purchasers that the Mayfield tractor was of robust construction yet light to handle and was suitable for lady operators. Another advertisement in 1959 advised that ploughing conserved the soil and

suggested that the Mayfield could also conserve the user's energy and in addition save both time and money. Fourteen different attachments were made including a plough, various cultivating implements and a front-attached cutter bar mower.

The first two-wheel Trusty tractors were assembled by Tractors (London) Ltd at Tottenham in 1933 from parts made by Walter Kiddy & Co., who were manufacturers of fire extinguishers. The company, founded by Mr J.C. Reach, moved to the White House at Bentley Heath near Barnet in 1938. Trusty was a household name in the two-wheel tractor world by the mid 1940s and by this time many implements including ploughs, cultivators, a disc harrow, ridger, hoe, mower, trailer and transplanter were being made at the White House works.

The ten-thousandth Trusty was made on 22 July 1947, when weekly output was between 80 to 100 tractors, and during the peak production period in the early 1950s about one hundred people were employed to meet orders from home and abroad.

Although short on styling, the Trusty tractor with its automatic centrifugal clutch, chain drive to a countershaft and from this two chains to drive the wheels with a dog clutch in each hub for power turning, was an extremely functional machine. Various petrol engines including Douglas, J.A.P and Norton were used during its long production run. Some of the later tractors had a Petter diesel engine; however, it was considered to be too heavy and was replaced with a Sachs diesel, which suffered from starting problems.

The Trusty did not have a gearbox and its top speed was 2 mph. A reverse gear was introduced in 1946 as a factory-fitted extra which cost £10. This was engaged with a deadman's handle to obviate any danger to the operator through stumbling or falling when reversing the tractor. Implements were hitched to a swinging drawbar or wheeled toolbar under the Trusty's unusually long handlebars, and a wheeled bogey seat was available for use with a cutter bar mower, transplanter, trailer, etc.

A radio-controlled Trusty using ex-RAF equipment was demonstrated to a large audience in 1946. A cylinder of compressed air carried on the tractor operated the controls and the event attracted wide press coverage. One magazine headline asked if armchair ploughing had arrived!

Plate 1.9   A reversible plough was one of many attachments made for the Trusty tractor.

In two minutes—

A lad can fit a plough

In two hours —

the lad can learn to use it

In two days —

he can plough two acres

ASK FOR A DEMONSTRATION

TRUSTY TRACTORS

FIRST IN THE FIELD
AND FIRST OUT

Fig 1.10   A 1947 sales poster for Trusty tractors.

An extensive 1947 price list for Trusty tractors and equipment included the Model 5 with a $4\frac{1}{2}$ hp J.A.P four-stroke engine on steel wheels for £125. The Model 6 with a $7\frac{1}{2}$ hp Douglas four-stroke engine or a $14\frac{1}{2}$ hp Norton engine was £130, pneumatic tyres being an extra £10. Every imaginable accessory was included from a one-way plough at £30, flat roller at £22 10s 0d and Trusty transplanter with easy-feed attachment for £45 to an engine cover at £2, a large grease gun for 7s 6d and a 10 gallon drum of Trusty engine oil for £3 15s 10d.

Trusty tractors hit the headlines again in 1948 when in an attempt to increase food production it was decided to plough up the wide verges on each side of the Barnet bypass. Five Trustys in line ploughed the verges, barley was sown and the sight of a tractor and binder harvesting the crop drew still more publicity.

The Trusty Imp, introduced in 1949, was a scaled-down version of the original Trusty with a $2\frac{1}{2}$ hp Villiers four-stroke engine, similar transmission system and a centrifugal clutch. The Imp, complete with an 8 in. digger plough and 3 ft light cultivator, cost £100. A 1950 sales leaflet pointed out that Trusty tractors led the way with ample reserves of the right type of power with a silky start, smooth running, self-steering and a specially designed tool for every job on the holding.

Plate 1.10   A Petter diesel engine was available for the Trusty.

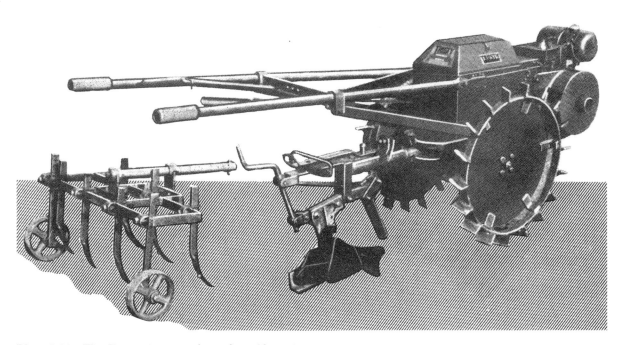

Plate 1.11   The Trusty Imp was introduced in 1949.

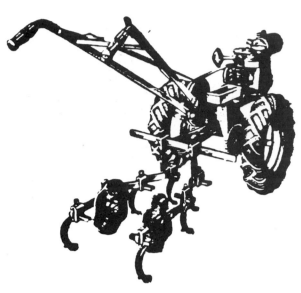

Fig 1.11   Vee-belt transmission on the
American-made Planet Junior B-1 provided a forward
speed of $1\frac{1}{2}$ to $3\frac{1}{2}$ mph.

The Trusty Earthquake rotary cultivator, with a wide range of attachments, was introduced in the mid 1950s. Buyers could choose a $2\frac{1}{2}$ hp Villiers Mark 20, 3 hp Mark 25 or a $4\frac{1}{2}$ hp Mark 40, air-cooled power units with a dry plate clutch and a three-speed gearbox with an optional reverse gear. The Earthquake had individual chain drive and reduction gear boxes to both wheels, and the three-speed cultivating rotor was driven by a heavy duty roller chain running in an oil bath.

Front-mounted cylinder, cutter bar and rotary mowers were made for the Earthquake, and other attachments included a plough, cultivator, ridger and trailer.

American-made two-wheel Planet Junior garden tractors were sold in Great Britain during the 1940s. The Planet A-1 and B-1 had 1 and $1\frac{1}{2}$ hp Briggs & Stratton engines respectively, while their big brothers the HT and B-HT were sold with either a 3 or $3\frac{1}{2}$ hp power unit. All were chain driven to pneumatic-tyred wheels, and track width adjustment between $14\frac{1}{2}$ and 20 in. was provided on the larger models. A gardening writer at the time advised readers that the Planet Junior garden tractor was well worth consideration and should be given a trial.

Plate 1.12   A hedge trimmer, chain saw and hand-held rotary cultivator were among the attachments that could be used with the flexible drive shaft on the Trusty Earthquake.

The Atco two-wheeled tractor, designed for market gardeners, smallholders and fruit growers, was said to be capable of doing the work of two horses and many other stationary tasks as well. An advertisement for this two-stroke-engined tractor gave little information but offered to send a catalogue dealing with the countless functions of this remarkable machine, which should be investigated by all who had any considerable extent of land or garden under their control. It concluded by stating that no money, effort or research had been spared in making the Atco tractor the finest of its kind yet produced.

Gravely Overseas at Buckfastleigh in Devon exhibited the Gravely Model D at the 1950 Smithfield Show. It cost £80 complete with tool frame and a tool kit containing a ring spanner, oil can, screwdriver, adjustable spanner and

Fig 1.12   A rare illustration of the Atco garden tractor, which was demonstrated to smallholders and market gardeners in the late 1930s.

Plate 1.13   The toolbar could be used in front or behind the single wheel of the Gravely Model D.

starting rope. Originally made in the 1930s by the Gravely Motor Plow and Cultivator Co. in West Virginia, USA, the cleated single-steel-wheeled Model D had a Gravely $2\frac{1}{2}$ hp four-stroke engine. It did not have a gearbox and the throttle was used to vary the forward speed between 1 and 3 mph. Attachments for the tool frame, which could be used in front of or behind the tractor, included cultivator tines, furrower, hoe blades and a plough.

The Gravely Model L two-wheel garden tractor, also sold in Great Britain during the 1950s, was designed for estate owners and market gardeners. It had a 5 hp air-cooled engine, two forward and two reverse gears, differential steering and an optional belt pulley. Numerous front-mounted attachments were available including a rotary plough, cultivator, roller, lawn mower, snow plough, saw, sprayer and rotary brush. The four-bladed rotary plough, protected by a slip clutch turned at 800 rpm, ploughed a furrow 7 in. deep and 12 in. wide in light soil, narrower in hard conditions.

Garner Mobile Equipment of London, who

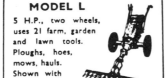
Fig 1.13   Gravely Model D and Model L garden tractors were made in America.

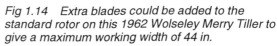

Fig 1.14   Extra blades could be added to the standard rotor on this 1962 Wolseley Merry Tiller to give a maximum working width of 44 in.

Plate 1.14   A top speed of 5 mph was possible with the 6 hp Coleby Senior. Axle extension shafts were used to set the wheel track between 15½ in. and 36 in.

were making garden tractors as far back as 1919, introduced the two-wheel Garner Light Tractor in 1947. The specification included a 5 hp J.A.P petrol engine with an optional tvo conversion kit, twist grip throttle on the handlebars and a centrifugal clutch. The three forward and reverse gearbox provided speeds between ½ and 8 mph; power was transmitted to the wheels through a differential with roller chain and sprocket drive from both half shafts to the wheels. Independent expanding shoe brakes on the half shafts provided power steering. The standard Garner tractor on steel wheels cost · £139 15s 0d; pneumatic wheels were an extra £7.

The Garner De-luxe model, priced at £157, included axle extensions, wide spade lug wheels, which could be split when working between narrow rows, extra balance weights

and an engine-mounted power take-off driven belt pulley. A floating toolbar made in collaboration with Stanhay, an off-set Wilmot plough, which enabled the user to walk on undisturbed ground instead of in the furrow, a trailer and bogie seat were some of the implements made for the Garner.

Coleby Cultivators of Swanley in Kent made the Senior and Junior garden cultivators and the Minor motor hoe in the 1950s. The Senior row-crop cultivator had a 6 hp Coleby engine, hand-lever-controlled friction plate clutch, three forward gears and reverse, wheel clutches for power steering and adjustable wheel track. The handlebars could be moved to either side of the cultivator, allowing the user to walk by the side of the machine. Sales literature advised that fuel consumption was one gallon per five hours cultivating and continued that the machine was so easy to manoeuvre that it could be handled by either sex with very little tuition.

The Coleby Junior row-crop cultivator with a J.A.P $3\frac{1}{2}$ hp engine was said to run for seven hours on one gallon of petrol. Forward speed varied between $\frac{1}{2}$ and $1\frac{1}{2}$ mph depending on throttle setting, and steering clutches aided turning on the headland. The Senior and Junior were supplied with front and rear toolbars, tines and hoe blades with a plough, seed drill, and cutter bar available at extra cost.

The Wolseley Sheep Shearing Co. of Birmingham were well known for their farm equipment, including stationary engines, elevators, sheep shearing machines and electric fencers, long before their involvement in garden machinery. The company changed its name to Wolseley Engineering in 1955 and obtained a licence from the Merry Manufacturing Co. in America to manufacture the Merry Tiller garden cultivator in 1957.

Wolseley merged with Hughes Engineering in 1958 to form the Wolseley Hughes Group. Five years later H.C. Webb Ltd accepted an offer to join Wolseley Hughes, who then acquired control of the lawn mower business but Webbs continued to trade as a separate concern. H.C. Webb changed its name to Webb Lawnmowers in the late 1960s and by the early 1970s a large part of the Wolseley factory was making Webb machines. Demand for lawn mowers was increasing, so the two companies decided to integrate and Wolseley Webb was formed in 1973. The Merry Tiller was made throughout this period and was still in production when Qualcast acquired Wolseley Webb in 1984.

Plate 1.16   The Merry Tiller Titan could be converted from a rotary cultivator to an inter-row hoe in a matter of minutes.

Two models of the Merry Tiller were made in 1959 with 1 hp and $2\frac{1}{2}$ hp engines, costing £57 and £59 respectively. The cultivating rotor was standard equipment and there were many accessories including a toolbar, trailer, cylinder lawn mower, cutter bar, saw bench and flexible drive shaft for a hedge trimmer and sheep shearing equipment. A pair of pneumatic wheels replaced the cultivating rotor when the Merry Tiller was used with this equipment.

The Merry Tiller Major, with a $2\frac{1}{2}$ hp Briggs & Stratton lightweight aluminium four-stroke engine, and the Professional, with a heavy duty

Plate 1.15   Ideal for small or large gardens, the 1960 Merry Tiller Major with a $2\frac{1}{2}$ hp Briggs & Stratton engine could be used with numerous row-crop tools and other attachments.

$2\frac{1}{2}$ hp Clinton engine, were made in the 1960s. Both had a recoil starter with a vee-belt tensioner pulley used as a clutch to engage the chain and sprocket drive to the rotor shaft.

The 1965 Merry Tiller range included the 5 hp Titan with various attachments, one of these being an unusual reversible plough of similar shape to a ridging body with slatted mouldboards said to turn a furrow from 6 in. to 10 in. wide and deep. The 5 hp Briggs & Stratton-engined Titan GT, with a two forward and reverse gearbox, was added in 1971. A two-speed vee-belt drive doubled the gearbox ratios to give the cultivating rotor four forward speeds, from 33 to 150 rpm, and 37 or 85 rpm in reverse. The Titan GT rotor was gear driven, which differed from the vee-belt arrangement on earlier models.

Wolseley entered the small garden cultivator market in 1971 with their 2 hp four-stroke Briggs & Stratton powered Trident with a 3 tine vertical rotor on an extended shaft in front of the engine. The Trident was trailed when digging so that no footmarks were left on the freshly cultivated ground.

The 1975 Wolseley Webb catalogue included the 3 hp Merry Tiller Major and Super Major, the two-speed 5 hp Titan and the 5 hp Titan GT, with four forward and two reverse gears. Within a year the Merry Tiller Major had a new 4 hp engine and the Super Major was upgraded to 5 hp. The Major cost £225 in 1976, the Super Major was £235, the Titan cost £295 and the Titan GT was £395.

The Wolseley Twin Six, made during the 1960s, could be used with up to six pairs of heavy duty slashers to give working widths from 3 ft to a massive 6 ft. A 9 hp Briggs & Stratton engine provided the power to drive the rotor through a two forward speed and reverse gearbox and vee-belt drive to the transmission case. Pneumatic tyred wheels were fitted on the rotor shaft for ridging, inter-row work and transport. The Twin Six cost £470 in 1976.

Two models of the Wolseley Webb Wizard garden cultivator were made after H.C. Webb Ltd joined the Wolseley Hughes group in 1963. The 3 hp and 5 hp Wizards had Briggs & Stratton engines to drive the slasher rotors to a maximum working depth of 12 in. Complete with 26 in. wide slasher rotor and folding front-mounted transport wheel, similar to that on the Merry Tiller Cadet, the 3 hp Wizard cost £166.67 in 1978 and the 5 hp model was £190.67.

Fig 1.15   A maximum cultivating width of 6 ft was possible with the Wolseley Twin-Six.

Qualcast launched the Cultimatic Super and de luxe garden cultivators in 1977. The two-speed Cultimatic Super, with a 98 cc Suffolk engine, cost £149.95 and the de luxe version with a 4 hp Briggs & Stratton engine and reverse drive was £199.95. The Cultimatics were similar to the Merry Tiller with the engine above the rotor, swing up transport wheels and a toolbar with hoe blades, cultivator tines, etc. Atco announced the 3 hp standard and 5 hp de luxe garden cultivators in 1982, both had Briggs & Stratton engines and except for the green and black colour scheme they appeared to be a close relation of the gold and black Qualcast Cultimatic.

The Qualcast Cultimatic B66, with a 114 cc Suffolk engine, replaced the standard and de luxe Cultimatics in 1985. It was painted green and cost £309.99. The B66 was made for three years and when it was discontinued in 1988 the price had risen to £321.43.

Improved and re-styled Super Major, Spartan, Centaur, Titan and Titan GT Merry Tillers with 5 hp Briggs & Stratton engines and the 3 hp Cadet were announced in 1983. The Super Major, Centaur and Titan had reverse drive and the Titan GT was said to offer the ultimate in garden cultivators with a four forward and two reverse speed rotor shaft driven by a combined vee-belt and gear transmission system.

Plate 1.17   The Qualcast Cultimatic B66 rotary cultivator was made from 1985 to 1988.

The Merry Tiller range was gradually reduced in the late 1980s. The Titan was dropped in 1989, the Major, Super Major and Cadet were discontinued in 1990 and the last Merry Tiller Titans were made in January 1991.

G.D. Mountfield of Maidenhead was established in the early 1960s to manufacture lawn mowers and garden cultivators. The M 1 Cultivator, announced in 1967, was described as the ultimate in powered garden mechanisation. It had a 3 hp engine to drive various attachments, including a rotary cultivator and a lawn mower or tow a trailer. The mid 1970s Mountfield rotary cultivator was of a similar design with provision to convert it into a grass cutter. The dual purpose power unit for a rotary cultivator and lawn mower was still made by Mountfield in the early 1980s. The $3\frac{1}{2}$ hp Briggs & Stratton four-stroke engine on the M 1 gardener could be used on a rotary cultivator unit or an 18 in. lawn mower. The 4 hp Mountfield Estate and 5 hp Monarch, both with Briggs & Stratton engines, were rotary cultivators with working widths from 9 to 36 in. A toolbar with hoes, cultivator tines and other implements was an optional extra for the three machines.

Mountfield was bought by Ransomes in 1985 and production of rotary cultivators has continued to the present day. The 24 in. wide Manor

Fig 1.17   A mid 1970s Mountfield rotary cultivator.

cultivator with the option of a 3 or 5 hp Briggs & Stratton engine was included in the 1994 Mountfield catalogue.

Westwood Engineering made Groundhog rotary cultivators at their Plymouth factory in the 1970s. The economy model of Groundhog G99 had a 3 hp Briggs & Stratton engine with vee-belt and roller chain drive to a 26 in. cultivator rotor. The heavy duty G4 and G5 were supplied with 4 and 5 hp engines.

Flymo added garden cultivators to their products in 1978. The Flymo GM with a 2 hp Ducati two-stroke engine had a 14 in. tiller rotor and the 19 in. rotor on the DM was driven by a 3 hp Briggs & Stratton four-stroke.

The range was expanded in 1980 with the purchase of Westwood garden cultivator designs. These machines were made, with some modifications, by Westwood Engineering exclusively for Flymo, who sold them in their orange and brown colours until 1984 when the complete range was discontinued. Built on the Merry Tiller style, two 3 hp and three 5 hp

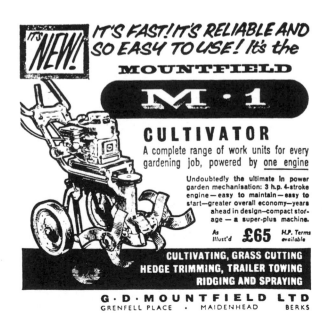

Fig 1.16   The Mountfield Model M1.

models were listed in 1982 with prices from a modest £156 to £313 for the Flymo GTM. A single-speed drive was standard on all models except the GLM, with a reverse gear, and the GTM, with five forward speeds and reverse.

A.J. Emery and Son of Halesowen exhibited their Universal rotary cultivator at the 1950 Smithfield Show. It had a 3 hp four-stroke, dry sump engine with an oil pump to ensure adequate lubrication no matter at what angle the machine was used. The Universal cultivator had a 3 speed gearbox with an optional reverse and separate clutches engaged the drive to the 14 in. cultivating rotor and the wheels.

Rototillers, made by George Monro at Waltham Cross in the 1950s, had their roots in

Plate 1.18   Emery Universal rotary cultivator.

Plate 1.19   The 8 hp two-stroke engined Rototiller 56 had a two-speed gearbox and could cultivate one acre in eight hours when used in low gear.

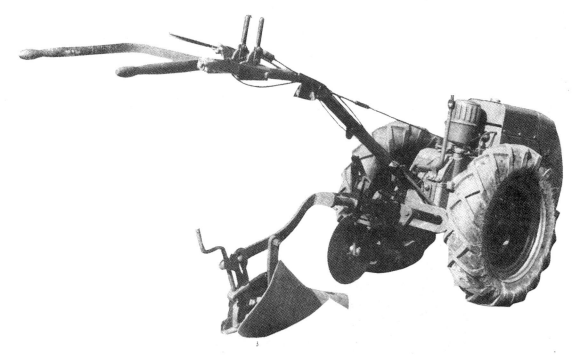

Plate 1.20   About two gallons of petrol was enough fuel to keep the Rowtrac model 5 ploughing all day.

Fig 1.18   The 1953 Monrotiller series II garden tractor.

the Swiss-made Simar Rototiller. The 1952 Monro range, all with two-stroke engines, consisted of the 5 hp Rowtrac garden tractor, Rototiller 35, 8 hp Rototiller 56 and $1\frac{1}{4}$ hp Monrotiller rotary cultivators. The Monrotiller was also made with a 1 hp four-stroke engine. The Series II Monrotiller with a 12 in. rotor, exhibited at the 1953 Smithfield Show, had a Villiers engine, pneumatic tyred wheels and complete with a tool kit it cost £80 ex works. Series IIc and III Monrotillers, with $2\frac{1}{4}$ hp overhead valve Villiers engines and a 14 in. rotor, were made in 1962. The Series IIc, with two forward speeds, cost £108 and the Series III, with three forward and three reverse gears, was £120.

The Rowtrac was a conventional two-wheeled garden tractor with a cone clutch, single forward speed and independent wheel clutches for power steering. Rowtrac sales literature stated that the fuel and maintenance costs for 8 hours ploughing was 6s 6d.

The Rototiller Minor, with a 98 cc two-stroke engine, made in the late 1950s, looked rather like an outboard motor with wheels and handlebars. It cost £59 and a small cultivating rotor with horizontal blades was included in the price.

An increasing number of imported garden cultivators appeared on the British market in the late 1950s. The Motohak, a wheelless rotary cultivator made by Klien Motoren in Germany, was tested by the NIAE in 1958. Power from the Solo engine was transmitted through a disc clutch, two-speed gearbox and chain drive to the rotor shaft. The rotor consisted of a series of cultivating stars and a maximum of six stars could be used on each half of the rotor shaft. The throttle, clutch lever and twist grip gear change were placed on the adjustable handlebars.

The Italian Motom M2 cultivator, with a 51 cc air cooled four-stroke engine, was imported by Trojan Agricultural Sales Ltd in the early 1960s. This small rotary cultivator could be used as a tractor by replacing the rotor with a pair of steel or pneumatic wheels.

Prospective buyers of a small garden cultivator in the early 1950s were spoilt for choice. The smaller machines made by Rotary Hoes and B.M.B were well established but they had to face competition from the Farmers' Boy, Troy Tractivator, Coleby and Anzani motor hoes, Barford Atom, etc.

The Farmers' Boy motor cultivator and light tractor, with a 1.2 hp J.A.P engine and dog clutch to engage drive to its steel wheels, was introduced by G.W. Wilkin Ltd, Kingston-on-Thames, in May 1950. Complete with toolbar, hoe blades, cultivator tines, power take-off

Plate 1.22 The Mark I Troy Tractivator, on pneumatic tyres with a 98 cc Villiers engine, cost £54 15s 0d in 1952.

pulley and engine tool kit, it cost £52 ex works.

Six years later the price of the Series II Farmers' Boy manufactured by Raven Engineering at Kingston-on-Thames had increased to £70. The specification included a 1½ hp Villiers Mark 10 air-cooled four-stroke engine, 14 × 3 heavy duty tyres, power take-off pulley and tool kit. A 10 in. toolbar with side hoes and cultivator tines was an extra £3 0s 0d, a general purpose plough was £23 0s 0d and a 10 in. rotary cultivator with 8 blades was priced at £12 15s 0d.

Raven Engineering increased the range of Farmers' Boy tractors with the Minor in 1958, the Minorette in 1959 and the Majorette in 1962. The Farmers' Boy Minor rotary cultivator with a two-stroke Villiers engine cost £48 and the four-stroke version was an extra £10.

The two-speed Minorette, with the choice of a Briggs & Stratton or a Villiers engine to drive the rotary cultivator, cost £56. The more powerful Farmers' Boy Majorette had a Briggs & Stratton four-stroke engine with a three forward speed and reverse gearbox. Attachments for the Minorette and Majorette included a rotary mower, hedge trimmer and toolbar with tines and hoe blades.

Advon Engineers had taken over manufacture of Farmers' Boy light tractors by 1967 and nine years later the Minorette and Little Giant two-wheeled rotary cultivators were still in production. The 1976 Minorette had a 129 cc Briggs & Stratton four-stroke petrol engine with a main drive clutch and two-speed drive,

Plate 1.21 According to the sales leaflet no other single machine in the world would do all the jobs that the Farmers' Boy light tractor could do, with better results, while saving time and money.

PRICE LIST

1st October, 1956

THE "FARMERS' BOY"

Light Tractor and Cultivator

£ s. d.

STANDARD TRACTOR—Powered by 1¼ h.p. (120 c.c.) air-cooled 4-stroke Villiers Engine, and fitted with 14" x 3" Heavy Duty Tractor Tread Tyres, Power Take-off Pulley at engine speed and Tool Kit ... **£70**

STANDARD TRACTOR, as above, fitted with FREE WHEELS 74 0 0

10" TOOL-BAR with 6" Side Hoes, Cultivating Teeth, Standards, and "U" Fixing Bolts, as complete set 3 0 0

ATTACHMENTS AND ACCESSORIES

GENERAL PURPOSE PLOUGH—STANDARD MODEL—complete with stepped axle, 22" x 4" Cleated Cast Iron Wheels and 10" fully adjustable Disc Coulter 23 0 0

GENERAL PURPOSE PLOUGH—HEAVY DUTY FREE-WHEEL MODEL—complete with Stepped Axle, heavy duty cleats and chains large mouldboard and landside with 10" fully adjustable Disc Coulter ... 32 16 0

10" ROTARY HOE fitted with eight blades 12 15 0

ROTARY SYCTHE—16" cutting disc 11 10 0

18" ROLLER LAWN MOWER complete with GRASS BOX 35 14 6
(plus P.T. £7 . 6 . 8)

3 ft. CUTTER BAR 23 15 0

"CLIPPGEARS" Hydraulic HEDGE TRIMMER
MODEL A.1 for fitting to "Farmers' Boy" complete ... each 40 15 0
MODEL A.2 self-contained Unit fitted with Villiers Mark 10 engine complete... ... ... each 64 10 0

TOWING TRAILER with 16" x 4" pneumatic tyres and metal seat, 8 cwt. capacity 23 10 0

SAW BENCH complete—fitted with 12" circular Saw blade 18 10 0

8" DISC HARROW—in two banks of three—adjustable ... per set 6 13 0

SEEDER "Jalo Jiffy" pattern, with fixing bracket - Single Wheel Model 4 3 6
Twin Wheel Model 5 5 6

10" TOOL BAR 12 6

Fig 1.19   Farmers' Boy price list for 1962.

providing ½ to 1 mph in low and 1 to 2½ mph in high gear. The Little Giant with a 4 hp Briggs & Stratton power unit, clutch and two forward gears had a maximum speed of 1½ mph in low gear and 3 mph in high gear.

The Mark I Troy Tractivator made by Troy Agricultural Utilities in the late 1940s, could hoe, drill, ridge, spray and mow with its 1¼ hp Villiers or J.A.P four-stroke engine geared down to give forward speeds of 1¼ or 2½ mph. It may have been

coincidence but the Mark I Tractivator had a striking resemblance to the 1950 Farmers' Boy. Troy Agricultural Utilities had moved to Surbiton by 1952 and by this time the Mark II Tractivator with a 1½ hp Villiers engine was made alongside the Mark I machine. Additional implements suitable for both models included a spike tooth harrow, disc harrow, potato lifter, rotary scythe and saw, roller and trailer. The Tractivator's drawbar pull could be increased with a

Fig 1.20   A 1952 vintage Barford Atom 15.

Plate 1.23   The Barford Atom 15 with a 7½ cwt capacity trailer and a seat for the driver. The pneumatic-tyred wheels were fitted with internal expanding brakes.

front weight box, which was filled with concrete or other heavy material.

Sales literature declared that the Troy Tractivator was the World's foremost light tractor and cultivator and that in price, performance, versatility and reliability the 1½ hp Troy had no equal. It was pointed out that the tractor was economically just the job for all work on one to twenty acres, it did not require skilled labour for maintenance and operation and would never need to be idle.

The Barford Atom tractor, with a range of implements, was made by Barford Agricultural Ltd at Grantham in the late 1940s. It had a 1¼ hp Villiers engine with a fuel consumption of well under one pint per hour. Power was transmitted through a reduction gearbox and roller chain to the land wheels and power take-off shaft with a speed range of 300 to 600 rpm. There were two models; the 'single wheeled' Atom, with two

pneumatic-tyred wheels side by side under the engine, cost £62 10s 0d in 1950 and the two-wheel model was £69 10s 0d. A kit to convert the Atom from one to two wheel mode was an extra £7.

The Barford Atom 15, with a 1½ hp Villiers engine and vee-belt drive to an enclosed reduction gearbox, replaced the original tractor in 1951. It had two forward gears with top speeds of 1½ mph and 3 mph.

The 1959 Chelsea Flower Show provided the gardening public with their first sight of the Barford Atom 30, which could be used for ploughing, rotary cultivating, hoeing, grass cutting, etc. There were two engine options, either a 3 hp Villiers Mark 15 or a 4½ hp Clinton. Power was transmitted through a two-speed gearbox and vee-belt tensioned, with a lever operated jockey pulley to engage drive to the wheels and rotary cultivator.

Dashwood Engineering, Empire Works, London made the Colwood Model B Garden tractor and the Model RA rotary hoe in the late 1940s. Both had a Villiers Mark 10 1.2 hp four-stroke engine with a hand-start lever; the Model RA had three forward speeds and the Model B had a two-speed gearbox. Various tool bar attachments were produced for both models and it was an easy task to remove the rotor from the Model RA when preparing it for toolbar work.

Landmaster, who acquired Colwood garden tractors in the mid 1950s, made a Mark II Model B with a 1½ hp Villiers Mark 12 four-stroke petrol engine, said to use four pints of petrol in an average eight-hour day. A roller chain transmitted drive from a two-speed gearbox to a single land wheel for row-crop work, rotary cultivating, grass cutting and hedge trimming.

The Shay Rotogardener was another easily handled rotary cultivator for the domestic gardener. Introduced by J.E. Shay at Basingstoke in 1954, the digging rotor was driven by a 120 cc two-stroke engine and the fuel tank, which held enough fuel and oil mixture for about three hours work, was attached to the handlebars. The Rotogardener cost £49 10s 0d and a pair of rubber-tyred driving wheels to replace the rotor when using it with a toolframe cost £2 18s 6d. The handlebars could be off-set to either side on the less powerful 80 cc two-stroke engined Rotogardener 80 introduced in 1959.

The Versatiller (figure 5.16) was another small rotary cultivator for the vegetable garden, it was made by Farmfitters of Gerrards Cross in the late 1950s and their advertisements claimed that the multi-purpose Versatiller brought power gardening to every home at a reasonable price.

An Australian engineer, Arthur Clifford Howard, established a company called Austral Auto-Cultivators in 1921 to manufacture his own design of self-propelled rotary hoe. Within a year the first machine, with five sections covering a width of 15 ft and driven by a 60 hp engine, was in production. Several were made and then Howard, prompted by the growing popularity of the Fordson tractor in Australia, adapted his rotary hoe for use with this tractor. His attention then turned to pedestrian-operated machines and by 1924 the first petrol-engined 6 hp Howard Junior rotary hoe was in use.

A.C. Howard then decided to build a tractor for his rotary hoe and the Howard Auto Rotary Hoe 22 Model DH with a water-cooled, four-cylinder overhead camshaft petrol engine

Plate 1.24   This Colwood Model B Mark II motor hoe was made in the mid 1950s by the Colwood Division of Landmaster Ltd, London SE20.

appeared in 1928. The DH22 was imported by J & F Howard of Bedford and shown at the 1930 Royal Show. Howards of Bedford were leading implement manufacturers at the time but this company had no connection with A.C. Howard. The Bedford firm went out of business in the early 1930s but manufacture of the DH22 continued in Australia.

Soon after this, A.C. Howard made his way to Great Britain and teamed up with Capt. E.N. Griffiths to form Rotary Hoes Ltd at East Horndon in Essex in 1938. Capt. Griffiths' connection with A.C. Howard began when he came across the Australian-designed rotary hoe in 1928 and decided to import one for trial purposes. The trial was successful, so the Captain arranged for Howards of Bedford to manufacture the rotary hoe and a total of about one hundred machines were produced for various tractors including Austin, Fordson and Case.

Amazingly the small Rotary Hoes company acquired the famous steam engine manufacturer John Fowler of Leeds in 1946 and then a year later after introducing his new FD22 crawler tractor, A.C. Howard made a considerable profit in selling the John Fowler business to Marshalls of Gainsborough. Although production of the FD22, which had a special drive arrangement for use with a rotary hoe, was short lived it was to be the forerunner of the Howard Platypus crawler made at Basildon from 1953 to 1956.

A.C. Howard and Capt. Griffiths introduced the new Rotehoe Gem as a replacement for

Plate 1.25   The Howard BJ engine was used on early models of the Series II Gem, manufactured by Rotary Hoes Ltd at East Horndon in Essex.

Plate 1.26 The Howard Gem Series IV, introduced in 1954, had a twin cylinder 9.8 hp petrol engine. A vaporising oil conversion kit was available.

the Australian-designed and East Horndon-built pedestrian-controlled Rotehoe 5, 6, 10 and 12 in 1940. Parts were hard to come by during the war years and in order to keep the permitted level of production going, various makes of motorcycle engine, often obtained from scrap dealers, were renovated and used for the new machines.

The Rotehoe Gem was the forerunner of the Howard Gem Series II introduced in 1947. Buyers could choose a Howard BJ (British Junior) or a J.A.P 6 hp petrol engine on early models of the Series II machine but later models were supplied with a J.A.P power unit. A dry sump with an oil tank next to the fuel tank was a feature of both engines and the new Gem had three forward and reverse gears with a single

Plate 1.27 A modified Howard Gem with a Sachs two-stroke diesel and steering brakes, working in a Thailand paddy field in 1957.

Plate 1.28   The Howard Bantam, with a Villiers engine.

plate dry clutch controlled by a lever on the handlebars. A 20 in. rotor was standard but 18 and 24 in. rotors were also available. Production peaked at 200 per week in the early 1950s.

Rotary Hoes Ltd was originally established at East Horndon but a decision by the local authority to alter the name to West Horndon in the early 1950s meant that the company changed their address but remained in the same premises!

The Series III and IV Gem Rotavators replaced the previous model in 1952, the Series III had a 6 hp four-stroke petrol engine to drive the land wheels and either an 18 or 20 in. rotor. The more powerful Series IV used a 9.8 hp twin cylinder four-stroke unit to drive its 20, 24 or 30 in. rotor at 172 rpm.

According to Gem Series III and IV sales literature, these machines were just the job for the smallholder because they had car type controls with the gear lever in a four-star quadrant on the handlebars, which were adjustable for height and lateral movement, and a differential lock, which automatically disengaged when the rotor drive was in neutral. A Sachs diesel engine was added to the options for the Gem Series IV in 1957. The single-cylinder two-stroke water-cooled engine developed 9 hp at 2,000 rpm and was said to work for eight hours on a gallon of fuel. Although the chain-driven rotor was a permanent fixture, other attachments including a furrower and roller could be hitched to the rotor cover and the Gem's power take-off pulley could be used to drive a soil shredder.

The 1.95 hp two-stroke Villiers-engined Howard Bantam, with four forward speeds, made its debut in 1950. Controls on the laterally adjustable handlebars included a lever to select either of the two gears, a rotor engagement lever, a control rod to tilt the engine platform over centre for tensioning the vee-belt drive and a throttle lever. A high/low speed option was provided by twin vee-belt pulleys on the engine output shaft. The rotor was easily removed, allowing the use of several attachments including a toolbar, sprayer, seeder unit, hedge trimmer, cutter bar and cylinder mower.

A Howard price list dated June 1956 included the Bantam with a 10 in. rotor at £99 0s 0d or £49 10s 0d down plus twelve monthly payments of £4 10s 10d. There was a choice of three engines, a 1.9 hp Villiers Mark 15 or B.S.A. four-stroke and a Villiers Mark 25c 2 hp two-stroke.

The Bullfinch and Yeoman joined the Rotavator family in 1955. The Bullfinch was initially sold as the Howard Bulldog but the name had to be changed following a dispute with Lanz, the German tractor manufacturer. It had a 1.25 hp four-stroke J.A.P engine with separate vee-belt

drives to the wheels and the 10 in. rotor. A Howard sales leaflet suggested that not so long ago, mechanised gardening was a luxury for the few but the Bullfinch had put freedom from the spade and fork within the reach of all.

The Yeoman completed the mid 1950s Rotavator range. It was supplied with a 4.2 hp Villiers or a 4.5 hp B.S.A. four-stroke engine. The transmission consisted of a single plate clutch and a gearbox with two forward speeds and reverse doubled up with a high/low ratio vee-belt drive. The single lever quick release clamping system used on the Bantam secured the 15 in. rotor unit and other attachments

Plate 1.29   The Bullfinch was the smallest member of the Howard Rotavator family in the mid 1950s.

Plate 1.30 Wash and brush up time for this Howard Yeoman Rotavator after a hard day on the allotment!

including front and rear toolbars, cutter bar, sprayer and seeder unit.

Clifford Cultivators Ltd were also making rotary cultivators in the late 1940s and during a ten-year period from 1949 the combined total sales of garden rotary cultivators by Rotary Hoes and Clifford exceeded 150,000 units. Clifford sales leaflets issued in 1954 gave the company address as West Horndon and in 1957 Rotary Hoes bought out the rights to the Clifford range and factory tooling while Clifford continued in the aircraft business. Two years later an improved range of Howard-Clifford garden Rotavators was launched at the 1959 Smithfield Show.

The Howard Rotavator Co. continued to sell trailed and mounted tractor Rotavators to farmers while the Howard–Clifford badge was used on machines built for gardens and smallholdings but within three years 'Clifford' was dropped from the logo. The word *Rotavator* is a palindrome (it reads the same forwards or backwards) and has become the standard term for rotary cultivators in much the same way as JCB is synonymous with industrial diggers.

The 1960 Howard–Clifford price list included the Bullfinch at £73 7s 6d, the Bantam was £104 and the Gem retailed at £240. The Demon, a more powerful version of the Bullfinch with a

Plate 1.31 The German-made Hako rotary cultivator, made in several working widths, was sold by the Howard Rotavator Co during the mid 1950s. This 12 in. model cost £73 7s 6d in 1959.

Plate 1.32   A handful of Howard Gem 800 prototypes were built in 1965 but it never went into production. It had a twin dry plate clutch, four forward and reverse gears and a top speed of 13 mph.

2 hp Villiers engine, cost £81 10s 0d and two new models, the Howard–Clifford 400 cost £175 and the 700 was £215.

The Howard Rotavator production facilities were expanded in 1960 with additional premises at Harleston in Norfolk followed by a new forge and foundry at Halesworth in 1962. The outgrown West Horndon factory closed in 1974 with the opening of a new assembly plant at Ipswich and component manufacturing facilities at Washington, Tyne and Wear. The closure coincided with a change of name from Howard · Rotavator Co. to Howard Machinery Ltd together with a new logo consisting of four Rotavator blades arranged to form a letter H. A new UK head office and spare parts depot was established at Saxham near Bury St. Edmunds. This was closed in 1980 and the head office moved to Harleston.

The 400 was originally built in the late 1950s to meet a demand from American rental yards (tool hire centres). Early models had a 5.5 hp J.A.P four-stroke petrol engine but by 1967 it had been replaced with a 5.25 hp Kohler K 161 power unit. The more powerful 700 was sold with a Villiers 7 hp governor-controlled two-stroke petrol engine or a 7 hp two-stroke German-made Hirth diesel. The 700 specification included the Snaplock quick release implement attachment system previously used on the Yeoman and the Selectaspeed gearbox in the transmission provided forward speeds from 0.7 to 10.4 mph.

The Howard Rotavator 300 introduced in 1961 had a 3 hp Kohler engine but it was generally considered to be underpowered and was replaced in 1965 by the 350 with a Kohler 5.3 hp engine, which had an automatic decompression mechanism for easy starting.

The 200, an improved version of the Bantam,

Plate 1.33   A 5.3 hp Kohler engine provided the power for this Howard Rotavator 350. Drive to the wheels was engaged with a cable-operated cone clutch.

Plate 1.34   The standard rotor on the 2 hp Howard-Clifford Demon could dig to a maximum depth of 7 in.

made its debut in the mid 1960s. A 2 hp Villiers 12/1 four-stroke engine provided the power and like the Bantam it had a vee-belt and worm drive to the rotor and wheels.

More than 50,000 Gems had been sold by the time the Series V arrived in 1960. A Howard 810 cc twin cylinder four-stroke petrol engine developed 12 hp to give the Series V with a twin plate dry clutch a top speed of 2.8 mph from its three forward and reverse gearbox. The 172 rpm rotor with a choice of 20, 24 or 30 in. working widths could cultivate to a maximum depth of 9 in.

There was a choice of a 10 hp Hatz ES 780 diesel or a Kohler K301T single cylinder petrol engine for the Series V Gem by the early 1970s. Further improvements were made in 1976 with the introduction of the Standard Gem and Super Gem.

The Standard Gem was little changed except for a single-plate clutch and a higher top gear but the Super Gem was beefed up with strengthened rotor shields and more powerful engine options. According to sales literature the Super Gem would be able 'to stand up to the most brutal use'. The extra power for the Super Gem came from a 15 hp Kohler four-stroke petrol engine or an 11 hp Hatz four-stroke diesel.

The 220 and 350 were among the family of Rotavators during the mid 1970s. Few changes had been made to the 350 but the smaller 220, which started life as the Bantam and then became the 200, had a much improved specification including a 3.3 hp Aspera engine, clutch, diff-lock and handlebars adjustable for height and lateral swing. A working width of 15 in. was achieved by adding extra blades to the standard 10 in. rotor.

The Howard Dragon (Colour Plate 14), described in a sales leaflet as a two-wheeled tractor power unit with a range of easy-to-attach implements, appeared in 1979. The 8 hp Kohler petrol engine transmitted drive through a two-speed vee-belt to a gearbox with two forward speeds and reverse. There were two power take-off shafts, one to drive the 16, 24 or 36 in. Rotavator and the other was used for a rotary or cutter bar mower attachment. Other Dragon implements included conventional and reversible ploughs, furrower and trailer.

The 352 replaced the 350 in 1981, it had a 5 hp Briggs & Stratton engine with a cast iron cylinder and the angular engine cover above the

Plate 1.35   The Gem became the Dowdeswell 650 in 1985.

familiar Rotavator orange chassis was painted'
olive green. A 1983 price list included the 352 at
£800, the Dragon cost £1,500 and prices for the
Gem were from £1,750 to £2,538.

Howard Machinery Ltd continued operations
until 1985, when receivers were called in. This
led to the purchase of the UK premises and some
of the products by Dowdeswell Engineering.
The Howard name and remaining product lines
were bought by Farmhand and after trading
for a while as Howard Farmhand the company
became Howard G.B.

Manufacture of the 352 continued with a
Dowdeswell badge but by this time a 5 hp MAG
engine was providing the power. Dowdeswell
also carried on with the Gem but they were not
able to use the name that had appeared on
Howard machines since the Rotehoe Gem was
introduced in 1938. Four models of the Gem,
renamed Dowdeswell 650 pedestrian-controlled
rotary cultivator, were available in 1985. There

Plate 1.36   The Danarm Texas rotary cultivator.

Plate 1.37   The Kubota T350 Power Tiller with a ridging body.

was a choice of 20 and 24 in. rotors and a petrol or diesel engine were available in 1985. Prices bore no comparison with the early days of the Gem; they started at £2,559 for a 20 in. 650 with an 11 hp Kohler petrol engine to £3,361 for the 24 in. machine powered by a 12 hp Ruggerini diesel engine. A 30 in. Dowdeswell 650 with either a 16 hp Kohler petrol engine or a 13 hp Hatz diesel was added in 1987.

The Danish-made Danarm de luxe cultivator with a 3 hp Briggs & Stratton engine, two sets of slasher blades and twin adjustable rear wheels priced at £96 in 1972 was similar to the Merry Tiller. By the late 1970s five models of Danarm Texas de luxe cultivators with 3 or 5 hp Briggs & Stratton engines were imported. Prices ranged from £199 for the basic 3 hp machine to £297 for the 5 hp de luxe TD5 cultivator with two forward and reverse speeds. Attachments included a toolbar, cylinder and cutter bar mowers, rotary brush and lawn rake. Kawasaki engines were used on some of the Texas cultivators including the 5 hp 521 and the 7 hp 721 in the mid 1980s.

Norlett cultivators, also from Scandinavia,

Plate 1.38 **Electric starting was provided on the Kubota AT 60 Tiller.**

Power Tiller, designed for the professional grower, had a $4\frac{1}{2}$ hp engine, six forward and two reverse gears and complete with a rotor extension kit it cost £484. The single-speed T250 and the two forward and reverse T350 had $3\frac{1}{4}$ hp Kubota engines. The wheel-driven petrol engined AT60, AT70 and AD70 diesel with a forward and reverse rotor and electric starting were added in 1988.

The Japanese-built Iseki KT600R Tiller in green livery with a four-stroke side valve engine and

appeared in Great Britain during the late 1970s. J.T. Lowe, who made hand-pushed Jalo hoes, cultivators and drills in the 1950s, were importing Norlett cultivators in the mid 1980s. There were single-speed 3 and 5 hp cultivators with Briggs & Stratton engines and a 5 hp Kawasaki-engined model with two forward speeds and reverse. The Norlett 6000 with a 5 hp Briggs & Stratton engine, also with two forward and reverse gears, was imported by J.T. Lowe in 1988.

Japanese garden cultivators were not a threat to British manufacturers until the late 1970s when Kubota, Honda and Iseki became serious competitors in a declining market. The first Kubota walking tractors were made in the early 1960s but were not available in the UK until 1978 when six models of Kubota Power Tiller with $2\frac{1}{2}$ to $5\frac{1}{2}$ hp engines were imported by John Croft at Whitley Bridge in North Yorkshire.

The cheapest, single-speed Power Tiller T120 with a $2\frac{1}{2}$ hp engine and 18 in. rotor cost £214 and the $5\frac{1}{2}$ hp T720 was the most expensive at just over £600. It had six forward and two reverse gears, two-speed power take-off and steering clutches. In common with all Kubota Power Tillers, except the T120, it could also be used for ploughing, cultivating, ridging and grass cutting.

Kubota moved to Thame in 1982 when the new T250, T350 and T450 Power Tillers were introduced to the British market. The T450

Plate 1.39 **The $2\frac{1}{2}$ hp single speed Kubota T150 tiller could, with the handles folded, be carried in a car boot.**

ten forward gears and four reverse was introduced to British growers by Mitsui & Co., Denmark Street, London SE5 in 1967. Chain Saw Products at Manchester distributed Iseki tillers from 1974 to 1979, when a Lely-Iseki partnership was formed to market all Iseki equipment in the UK. The 1979 price list for Iseki Tillers, now painted red, included the 5 hp two-stroke AC20 and $3\frac{1}{2}$ hp four-stroke AC40 rotary cultivators, the 3 hp two-stroke KS280 garden tractor and the $7\frac{1}{2}$ hp KC450 two-wheel tractor with four forward and four reverse gears and power take-off.

The MC-1 mini rotary cultivator for small gardens had a two-stroke 2.2 hp engine above the 18 in. rotor with a flexible air inlet tube clipped to the handlebars to reduce dust intake.

Iseki UK was established when the Lely-Iseki partnership ended in 1986. The new company offered British growers a range of $3\frac{1}{2}$, 5, 7.3 and 11 hp Iseki tillers and garden tractors with a choice of rotor widths, gear ratios and implements. The 1988 Iseki catalogue included the AC20 and AC40 rotary tillers and A400 garden tractor but by 1993, although Iseki UK were marketing lawn mowers, tillers were no longer available.

By the late 1980s, with the exception of British-made Merry Tillers and Dowdeswell rotary cultivators, the UK market was dominated by imported machines including those made by Agria, BCS, Bertolini Ferarri, Gravely, Kubota and Roper.

Fig 1.21   The Iseki KT600R Tiller had ten forward and four reverse gears.

Plate 1.40   The 4 hp four-stroke Iseki KC450 power tiller was made during the late 1970s.

2 Ride-on Tractors

Market gardeners and smallholders often hired the services of local farmers to plough their land during the 1940s but farm tractors at the time were not really suitable for row-crop work and were too cumbersome to use in confined areas. However, by the late 1940s several makes of small ride-on tractor were being manufactured, and this meant that it was no longer necessary for market gardeners to spend long hours trudging behind a two-wheeled tractor.

Ransomes and Bristol crawler tractors were popular with some market gardeners in the late 1940s but the more versatile three- and four-wheeled 5 to 10 hp tractors including the B.M.B President, Byron, Garner, Gunsmith, Newman, OTA, Singer Monarch and Trusty Steed were sold in considerable numbers. The sales boom continued for a few years but the Ferguson 20 tractor, with its fingertip control hydraulic system, which cost just over £300, brought an equally rapid decline in sales of the President and its competitors. Small farm tractors met the needs of smallholders for several more years but later models became too big, and a new generation of small tractors imported from Japan filled the gap.

The three-wheeled ride-on Gunsmith Mark I light tractor, designed by Frederick Gunn and Harold Smith, was sold by Farm Facilities of Maidenhead. The Gunsmith cost £178 when it was introduced in 1948 but a later catalogue included the steel-wheeled tractor at £185, pneumatic tyres were an extra £12 10s 0d and with turf tyres it cost £212 10s 0d. The engine and two-speed transmission unit used on the B.M.B Plow-Mate was bought in for the tractor and like the Plow-Mate, most of the Garners had a 6 hp J.A.P engine but some were made with a Briggs & Stratton Model ZZ power unit. Twin vee-belt pulleys on the engine and gearbox input shafts with the drive belt moved by hand doubled the number of gears to give a top speed of 5 mph.

The driving seat positioned in front of engine gave an uninterrupted view of the mid-mounted toolbar, which like the plough and other implements, was raised and lowered with a hand lever. Handlebars were used to steer the single front wheel and pedal-operated differential brakes aided headland turns. Drive was engaged with a pedal-operated 'clutch', which tensioned the drive belt from the engine. Sales

Plate 2.1   The engine was behind the driver's back on the Gunsmith Mark I light tractor. This must have provided some comfort on cold winter days.

Fig 2.1   The steering wheel on the Mark II Gunsmith replaced the handlebars on the earlier model.

Plate 2.2   The steel-wheeled Garner cost £197 15s 0d when it was introduced in 1949. Pneumatic tyres were an extra £5.

literature pointed out that by merely replacing the belt, this type of clutch could be overhauled in a matter of seconds with no mechanical knowledge required.

A steering wheel replaced the handlebars on the Mark II Gunsmith, which appeared in the early 1950s, and a pair of multiple vee-belt pulleys with four grooves on the engine and gearbox output shafts gave a top speed of 8 mph.

Between 200 and 250 Gunsmith tractors were made during its four-year production run. A testimonial letter from a satisfied owner declared that the Gunsmith was wonderful, as it did five hours work in one hour while he sat in an armchair (tractor seat). Another user wrote that a boy of sixteen could do everything required with the tractor.

The four-wheel Garner light tractor, developed from the earlier two-wheel model, used the same 6 hp J.A.P Model 5 air-cooled petrol engine, centrifugal clutch, gearbox, brakes and final drive. The throttle control was conveniently placed on the steering column in front of the driver, who sat with his back to the engine. However, the gear change lever was in the same position as it was on the two-wheeled Garner, which meant that the driver had to reach behind the seat to change gear. Three forward gears gave a range of $\frac{1}{2}$ to 10 mph and wheel track was adjustable from 25 in. to 42 in. The Garner had a power take-off, belt pulley and rear drawbar for trailed implements. A mid-mounted toolbar and a rear-attached plough were raised and lowered with a hand lever.

The 6 hp four-wheel Garner was considered to be rather under-powered and this led to the introduction in 1950 of a slightly longer and wider model with a 7 hp J.A.P Model 6 petrol or paraffin engine which cost £269 15s 0d. A mid-mounted one-way plough attached in the same way as the underslung toolbar and the usual range of cultivation equipment, seeder units, potato and sugar beet lifters were made for the Garner light tractor. Production of Garner tractors came to an end in 1955.

Byron Farm Machinery of Walthamstow entered the small tractor market in 1947 with a three-wheel model designed for the smallholder and market gardener. The Byron Mark I, in common with similar tractors at the time, had an industrial version of the popular Ford 10 petrol engine. The combination of a single front-wheel and independent rear-wheel brakes not only gave excellent manoeuvrability but also helped keep the price at a competitive level. A mid-mounted toolbar was standard equipment and the Byron was claimed to be powerful enough to pull a two-furrow plough or a 5 ft cultivator in normal conditions.

A slightly improved Mark II version was

Plate 2.3 A diesel engine and hydraulic lift were features of the 1949 Newman WD2, which cost £410 12s 0d.

introduced in 1949, mainly recognised by its slimmed down mudguards, which improved visibility of work for the driver. The Mark II cost £247 10s 0d and the row-crop version was £292. When compared with £335 for the tvo engined Ferguson TED20 it was questionable whether the Byron offered value for money.

The Grantham area of Lincolnshire had its fair share of small tractor manufacturers in the late 1940s. The three-wheeled Kendall tractor (Colour Plate 1), made from 1945 to 1947, was one example. It had an 8 hp twin-cylinder air-cooled Douglas engine. The Kendall business was bought by Newman Industries at Bristol and production of the tricycle-wheeled Newman tractor started in 1948, with a choice of a $10\frac{1}{2}$ hp or a 12 hp Coventry Victor air-cooled petrol engine.

An improved Newman WD2, with a water-cooled single cylinder Coventry Victor diesel engine and a dry plate clutch, appeared in 1949. The conventional transmission with a three-forward speed and reverse gearbox and independent brakes gave $1\frac{1}{2}$, 3 and just under 9 mph at full throttle. The Newman diesel cost £330 in 1949 but with a belt pulley, power take-off and hydraulic lift the total price was £410 12s 0d.

The Newman Model E2 four-wheel tractor with a four-forward speed and reverse gearbox, giving a top speed of $8\frac{1}{4}$ mph, was introduced in 1951. Front wheel track was adjustable from 42 in. to 54 in. and the rear wheels could be set out to a maximum of 72 in. The E2 had a 12 hp Petter twin-cylinder water-cooled diesel engine, which was hand started with the aid of a decompressor and fuel priming lever. The basic tractor cost £430, extras included a rear belt pulley for £10, an $1\frac{1}{8}$ in. power take-off shaft for £18 and the hydraulic lift system, with a gear pump in the transmission housing, added £46 10s 0d to the price.

The Newman was supplied with a swinging drawbar and there was provision for front- and rear-mounted tool bars raised with a hand lever or an optional hydraulic system. High ground clearance made these tractors with a mid-mounted toolbar ideal for row-crop work.

Oak Tree Appliances of Coventry used the

Plate 2.4   Made by Oak Tree Appliances and marketed by Slough Estates, the 10 hp OTA tractor used three gallons of petrol in an eight-hour day.

Plate 2.5   The last Singer Monarch was made in 1956.

company initials for the OTA tricycle-wheeled tractor and following an initial appearance at a local agricultural event it was launched at the 1949 Smithfield Show. The first OTA tractors were red and yellow but later models were painted blue and this colour was used until 1955. Complete with hydraulic lift and 6 volt electric starting the OTA was priced more realistically than the Byron and Newman at £264 10s 0d. A four-speed power take-off with 10 in. diameter pulley was an extra £22 10s 0d and a set of wheel strakes was available for £11 6s 8d.

The OTA was another 1950s small tractor with a water-cooled four-cylinder side-valve Ford 10 hp industrial petrol engine with coil ignition. A Beccles vaporising oil conversion kit priced at £10 10s 0d was an optional extra. The three forward and reverse Ford gearbox was doubled up with a high/low range box to provide six forward speeds from $\frac{3}{4}$ to 15 mph and two reverse speeds. An unusual feature was the provision of a second starting handle dog on one of the shafts carrying the high/low range gears at the back of the tractor. When starting the engine at the rear it was necessary to leave the tractor in gear and set the high/low ratio lever in neutral.

The OTA had a worm and wheel final drive and a live hydraulic system with a twin-cylinder piston pump belt driven from the engine crankshaft. The chassis consisted of two steel channels set at an angle to raise the front of the tractor where a fork assembly carried the cable-steered pneumatic-tyred front wheel. This gave a tight turning circle and the resultant high-ground clearance provided an excellent view of the hydraulically operated underslung toolbar for row-crop work.

The redesigned bonnet and sheet metal radiator grille were the most obvious changes to the Mark II tractor, compared with the cast iron radiator grille with OTA in large letters on the previous model. Some of the Mark II tractors, known as the 5000 series, also had an improved transmission system and a power take-off shaft.

The four-wheel OTA Monarch Mark III was launched at the 1951 Smithfield Show. Apart from the obvious change to the front axle with an orthodox steering linkage and a front-hinged bonnet, the specification of the new tractor was almost identical to the tricycle version. Wheel-track settings were adjustable from 42 in. to 60 in. and various hydraulically operated implements including mid- and rear-mounted tool

Plate 2.6   Sales literature described the Trusty Steed as an easy to control little tractor, which could pull a man-sized load yet was easy to manoeuvre in a restricted space.

frames, ploughs, spike tooth and disc harrows, cultivators, potato lifter and a mid-mounted mower were made for it.

Production of the three-wheeled OTA came to an end early in 1953 when Singer Motors bought the manufacturing rights for the Monarch and started making the Singer Monarch Mark III at Birmingham. The change also terminated Slough Estates' marketing involvement with Monarch tractors. Singer made further improvements in 1955 with the introduction of the Monarch Mark IV. Apart from changing to orange paint the most significant modification was the inclusion of a standard three-point linkage. This was in line with the trend to standardise the dimensions of hydraulic lift linkages and mounted implements.

The last Monarch tractors were made at Birmingham in 1956 when Singer Motors was taken over by the Rootes Group. About 1,000 OTA and

Plate 2.7   About 500 Trusty Steeds were made. Some had an upright exhaust, bonnet and a power take-off for a Howard Rotavator and other power-driven equipment.

Monarch tractors were made between 1949 and 1956.

Tractors (London) Ltd. at Bentley Heath, already famous for their 6 hp two-wheeled garden tractors, introduced the Trusty Steed in 1948. It cost £200 and was basically a towing tractor with the Douglas engine used on the two-wheeled Trusty, which developed 8.9 hp at 3,500 rpm. A centrifugal clutch engaged drive to the rear wheels, the gearbox provided top speeds of 3½ and 8 mph and in addition to the normal front-wheel steering, a joystick control operated the power steering mechanism to give a tight turning circle.

An improved Trusty Steed, announced in 1950, cost £205. Unlike the earlier model it was suitable for row-crop work with a mid- or rear-mounted toolbar raised and lowered with a spring-assisted hand lever. There was a choice of a J.A.P or Norton engine, now mounted at the front of the tractor, which developed a maximum of 14½ hp. The new Steed had a single-plate clutch, three forward speeds and reverse, independent brakes, power take-off and adjustable wheel track. In addition to the usual range of toolbar equipment, many implements including

Plate 2.8   A half track conversion kit for the Trusty Steed became available in 1951. It was intended for use on marginal land and upland areas.

Plate 2.9   The Trusty steed with a Dorman sprayer.

a plough, mid-mounted cutter bar and trailer were made for the new Steed.

The Trusty Road Roller, two sizes of dumper truck (plate 6.37), a fork truck attachment and an angle dozer blade for the 14½ hp Steed were added to the product range in the mid 1950s and were made until Tractors (London) Ltd ceased trading in 1978.

A prototype of the B.M.B President, developed by Brockhouse Engineering at Southport, was exhibited at the 1950 Royal Show. Quantity manufacture started later that year and production continued for six years.

The President had an 8/10 hp Morris four-cylinder, petrol/paraffin engine with 6 volt electric starting, later changed to a 12 volt system.

Power was transmitted to the pneumatic-tyred wheels through a dry single-plate clutch, a three-forward speed and reverse gearbox, spur gear final drive and independent rear wheel brakes. Top speed was 8 mph and the wheel track was adjustable from 40 in. to 72 in. The basic price in 1951 with a swinging drawbar was £239 10s 0d but this had increased to £276 10s 0d by 1953.

Power take-off, belt pulley and the hydraulic lift, which was bolted to the side of the gearbox, were available at extra cost. The hydraulic unit was supplied with oil from the gearbox and two rams were used to operate a mid-mounted toolbar and three-point linkage. The mid-mounted toolbar had a simple depth control system with depth limiting pads on the frame, which rested

Plate 2.10   The B.M.B President was in production for six years.

Plate 2.11   The Stockhold President at work with a Hayter mower in 1958.

on the front axle when the hoes, etc. were in work. A screw handle adjustment on the limiting pads controlled the working depth of the hoes and when working on uneven surfaces the pivoting action of the front axle raised or lowered either side of the toolbar to maintain the correct setting.

The President was one of the more successful light four-wheel tractors but when vineyard and orchard versions were added in 1954 there was a fall in the demand for small tractors and the President was discontinued in 1956.

H.J. Stockton Ltd of London considered there was still a need for a small, light tractor that was little more than an engine, transmission, wheels, drawbar and a seat. This resulted in the launch of the Stockold President, based on its old B.M.B namesake, at the 1957 Smithfield Show. The new tractor was similar in appearance to the B.M.B President but had a two-cylinder air-cooled

14 hp Petter diesel engine with a three forward and reverse gearbox and pneumatic tyres. When announced it was expected that the price for the basic model, including a starting handle, would be about £360. Power take-off, hydraulics and electric starting were optional extras.

By 1960 there was an even more marked decline in the popularity of 6 to 10 hp four-wheeled tractors. This was due in no small way to the huge sales of Ferguson 20 and 35, Fordson Dexta, BMC Mini and similar tractors. However, small numbers of light tractors including the Martin-Markham Colt, Uni-Horse, Westwood, Wheel Horse and Winget were made during the 1960s and early 1970s.

Colt Tractors at Grantham built the four-wheel market garden tractors marketed by Martin-Markham from 1960 and about 200 were sold during its nine-year production run. There were two models, one with a 7 hp four-stroke, air-cooled Kohler engine, while the more power-ful de luxe Colt tractor was rated at 10 hp. Engine power was transmitted by flat belt through the clutch to a transfer box linked by a chain drive to a three-forward and reverse main gearbox on the rear axle. Hydraulic lift and power take-off were available and the trac-tor had only one rather ineffective rear-wheel brake. Implements for the Martin-Markham Colt included a plough, cultivator, loader, irrigation pump and tipping trailer.

The 7¾ hp Winget four-wheel light tractor was made by Slater and England, of concrete mixer fame, in the mid 1960s and most of the 500 or so units made at Gloucester were exported. The specification included a Lister SR1 552 cc direct injection diesel engine, a Newage gearbox and rear axle transmission unit, live hydraulics and power take-off. A front-end loader and rear-mounted implements including a plough, cul-tivator and transport box were made for the Winget tractor.

The first Uni-Horse garden tractors (Colour Plate 6) were made in Birmingham by Lea Francis cars in 1961 but this company eventually went into receivership. Production of these B.S.A. engined garden tractors was taken up by Uni-Horse Tractors Ltd and various improve-ments including wider mudguards were made in the late 1960s.

About 200 four-wheel Trojan Toractor garden tractors were also made during the early 1960s. A transverse 3¼ hp Clinton engine provided the power for the blue Mark I Toractor equipped with a set of pneumatic tyres. The red and white Mark III, which replaced the Mark I, had solid rubber front tyres and a conventionally mounted 4 hp Clinton engine.

Plate 2.12   A hand lever operated hydraulic pump on the right of the steering wheel raised the tool frame on the Rollo Croftmaster.

Fig 2.2   The Wheel Horse tractor was advertised in 1959.

The four-wheel Rollo Croftmaster, designed by a Scottish crofter, was made in various parts of Scotland for Rollo Industries at Bonnybridge in the late 1950s. There was a choice of 3 hp or 5 hp B.S.A. engines with the larger power unit intended for heavy work or work at high altitudes, which might cause a drop in power output.

The Rollo had a chain drive from the engine to a hand lever engaged multi-plate clutch and a three forward and reverse gearbox with a top speed of 6 mph. The pedal-operated single-expanding shoe brake on the differential unit was backed up with an additional hand lever, used when it was necessary to stop in a hurry.

A parallel-linkage tool frame, designed to keep implements level with the ground at all times, was lifted with a hand lever operated hydraulic pump with a separate oil reservoir. Six to eight strokes of the pump lever were said to be sufficient to raise the single-furrow plough well clear of the ground. A tap was used to release the oil in the cylinder and return the implement into work.

The 3 hp Croftmaster with hydraulic lift and tool frame cost £315 in 1957, the 5 hp model was an extra £13. Other prices included a single-furrow plough for £18 10s 0d, a 6 cwt two-wheeled trailer at £34 10s 0d and a 3 ft cut mower controlled from the driving seat was £66.

The basic specification of the International Cub Cadet tractor included a 7 hp Kohler engine with a recoil starter, three forward and reverse gearbox, dry single-plate clutch and pedal-operated band brakes on the rear wheels. Electric starting and hydraulic linkage, which simplified the operation of the normal lever-operated mid-mounted mower and rear tool frame were optional fittings.

The Ashfield 15-4E ride-on tractor made by Ashfield Agricultural Products at Derby was available with a 9, 11 or 16 hp Lister diesel engine. It had 12 volt electrics, four forward and reverse gearbox, hydraulic system and power take-off. The implement range included a loader, plough and cultivation equipment, rotary mower, transport box and tipping trailer.

Plate 2.13 The American- built International Cub Cadet was imported by the International Harvester Co. of Great Britain during the mid 1960s.

Plate 2.14   A mid 1970s Allen garden tractor.

Plate 2.15   The $12\frac{1}{2}$ hp Kubota B6000 compact tractor was launched on the British market in 1975.

Caplin Engineering of Ipswich made the Fieldrider 301 mini-tractor for a short period in the mid 1960s. Priced at £285 it had bright red fibreglass panels covering the Norton Villiers 147 cc four-stroke petrol engine and a complicated belt drive transmission with a four-speed and reverse transaxle.

By the early 1970s the trend in agricultural mechanisation was towards bigger and more powerful tractors and the development of small tractors for market gardens, nurseries, parks and orchards was virtually ignored. Large numbers of Japanese diesel-engined mini tractors were being sold in America and in 1974 the Marubeni Corporation of Japan decided to establish an outlet for the Kubota range of compact tractors at Whitley Bridge in Yorkshire. A range of matched equipment was already available for the 12½ hp Kubota B6000 tractor and implements were being developed for the L225 and L175 models. The first advertisement for Kubota compact tractors appeared in the British press midway through 1975.

The four-wheel drive Kubota B6000 had a two-cylinder water-cooled diesel engine with a six forward and two reverse gearbox, independent rear-wheel brakes and hydraulic lift.

The L175, with a two-cylinder 17 hp diesel engine with eight forward speeds and two reverse, two speed power take-off and category I hydraulic linkage, was next in the range. It cost just £1,770 in 1975 and the addition of a safety cab increased the price to just under £2,000.

The two-wheel drive L225 had a 24 hp three-cylinder water-cooled diesel engine with a gearbox, power take-off and hydraulic system similar to the L175. Complete with safety cab it cost a little over £3,000. A four-wheel drive version, the L245, appeared in 1977 and the three-cylinder 16 hp B7100 with six forward and two reverse gears and four-wheel drive was added in 1978.

More models were introduced in 1982 and having outgrown their Yorkshire premises Kubota moved to Thame in Oxfordshire. A wider selection of Kubota compact tractors was introduced to the British market but some of the more powerful models were not really suitable for horticultural work.

The Japanese Iseki Agricultural Machinery Manufacturing Co. was established in 1926 but Iseki compact tractors were not available in

Plate 2.16   The 24 hp L225 was the most powerful Kubota tractor available to British growers in 1975.

Plate 2.17   The 1988 Kubota four-wheel drive 12½ hp B4200 compact tractor cost just under £4,500.

Great Britain until 1976. This came about following a visit to Japan by Len Tuckwell, a well-known John Deere tractor distributor in Suffolk. He established a new company at Ipswich, trading as L. Toshi Ltd, to import Iseki TS2810, TS1500 and the four-wheel drive TX1300F (Colour Plate 8) compact tractors. The TS3510, 1910, 2110 and a two-wheel drive TX1300 were added a year or so later.

The TX13 and TX15 ranges had 13 hp or 15 hp two-cylinder Mitsubishi power units and a three-cylinder four-stroke water-cooled Isuzu diesel engine was used for the 28 hp TS2810 and the 35 hp TS3510. A four-speed power take-off, hydraulic lift and differential locks were standard equipment on the TX tractors. A nine forward and three reverse gearbox was fitted to the TS2810 and the TS3510 had eight forward and two reverse speeds. A two-wheel drive version of the TS1300F had six forward and two reverse gears and in common with all Iseki compact tractors at that time a safety start device

was built into the clutch pedal linkage.

The same models were being imported in 1979 when a Lely-Iseki partnership established at St. Neots in 1979 replaced L. Toshi Ltd as the distributor of Iseki compact tractors in the UK. This arrangement continued until 1986 when the partnership came to an end and Iseki UK was established, at Little Paxton near Huntingdon, and later moved to Bourn, also in Cambridgeshire . The TX range was still current with three-cylinder engines and an improved closed centre hydraulic system.

Seven years later in 1993, a marketing agreement between Massey Ferguson and Iseki resulted in MF farm tractors being sold in Japan with an Iseki badge and Japanese compact models marketed in Great Britain with an MF logo and red paint.

Massey Ferguson were not involved with garden tractors until 1965 when the MF10 Suburban Tractor was introduced in the United States and Canada. The 10 hp MF10 designed for

Plate 2.18   The TS3510 was one of the Japanese-built Iseki compact tractors imported by L.E. Toshi at Ipswich in the late 1970s.

small market gardens and the large domestic garden had a single-cylinder four-stroke air-cooled petrol engine. Gear changes were made on the move through a five-speed variable vee-belt drive unit combined with a four forward and reverse gearbox, which gave 20 forward and five reverse speeds. The 7 hp MF7 lawn tractor and the larger 12 hp MF12 with hydrostatic transmission were added a year or so later. A full range of cultivators, mowers and other implements including a trailer and snow blower were made for these tractors. They were sold in continental Europe, particularly France, and were made until an agreement with Toyosha of Japan in 1977 resulted in a new Massey Ferguson 1000 series of compact tractors.

Plate 2.19   Iseki TX 1300 two-wheel drive compact tractor.

Plate 2.20   The 16 hp Massey Ferguson 1010 was one of the Japanese-built 1000 series compact tractors introduced to the British market in 1984.

Plate 2.21   The 21 hp MF1020 Hydro with a rocker pedal and high/low gear lever to control the hydrostatic transmission was added to the Massey Ferguson compact tractor range in 1987.

The new tractors were also sold in parts of Europe but not in the UK. An MF1010 was exhibited at the 1983 Royal Show and this led to the introduction of the MF1010, 1020 and 1030 compact tractors to the British market in May 1984. All three were available with two- or four-wheel drive and had three-cylinder water-cooled diesel engines. The smallest 16 hp MF1010 with six forward and two reverse gears, power take-off and category I hydraulic linkage was suitable for the market gardener. The 21 hp 1020 and 27 hp 1030 had twelve forward and three reverse gearboxes with creeper speeds for specialist planting and harvesting operations.

An optional hydrostatic transmission with a rocker pedal to control forward and reverse travel and high/low ratio gear lever became available for the MF1010 and MF1020 in 1987. This simple change-on-the-move system was a far cry from the multiple vee-belt pulley gear change on the Gunsmith tractor in the early 1950s.

Clutchless forward and reverse speeds controlled with a single lever and four-wheel steering to give a tight turning circle were features of the MF30 Series lawn tractors introduced by Massey Ferguson in 1990. The three models designated MF30-13, MF30-15 and MF30-17 had hydrostatic transmissions and underslung side-discharge mower decks. The second pair of numbers indicated the horsepower developed by their Briggs & Stratton Vanguard V-twin petrol engines. A smaller MF20-12 with a single cylinder 12½ hp Briggs & Stratton engine and a six forward and reverse gearbox completed the 1990 range of MF lawn tractors.

The new 12 Series tractors were launched in 1993 by Massey Ferguson's grass equipment division under the agreement made with Iseki. Built in Japan, there are three compact models MF1210, MF1220 and MF1230 with 17, 20 and 25 hp diesel engines respectively also the 30 hp MF1250 and 35 hp MF1260 small tractors for golf courses, parks departments, etc. Options for the 12 Series include mechanical and hydrostatic transmissions, manually engaged front-wheel drive and power steering.

A dozen different models of crawler tractor in various sizes were made between 1945 and 1960 with the Bristol 20 and Ransomes MG of particular interest to smallholders and market gardeners in Great Britain and abroad.

The first Bristol crawler tractors with Douglas engines were made in 1933 at Bristol by an associate company of Roadless Traction Ltd. The Douglas engine was a 1200 cc horizontally opposed twin-cylinder air-cooled stationary engine adapted for automotive use. It was eventually replaced with a water-cooled Austin power unit. Engine speed was controlled with a twist-grip throttle on the single-tiller-style lever used to steer the tractor through independent differential brakes.

Publicity material described the Bristol crawler as a modern farm tractor which could be worked night and day at less cost than the upkeep of two horses. The Bristol was just under 3 ft wide and ran on 7 in. wide rubber jointed tracks developed by Roadless Traction. It weighed a ton, used one gallon of petrol an hour and cost £155 including delivery to the nearest railway station. Bristol Tractors were bought by Austin car distributor H.A. Saunders at Finchley in the mid 1940s and when production of the Bristol crawler resumed after the war, it had a four-cylinder 10 hp Austin engine suitable for petrol or tvo.

The Bristol 20 crawler introduced in 1948 had a modified 22 hp Austin 16 overhead valve car engine with magneto ignition and was only 3 in. wider than the previous tractor. Power was transmitted to the tracks through a single plate clutch, three forward and reverse gearbox and spur gear final drive units. The Roadless design of rubber jointed track was retained, steering was by multi-plate clutches controlled by two hand levers and independent foot brakes were

TAKES A UNIT-ATTACHED PLOUGH, TOOLBAR FRAME, ETC., AS WELL AS DRAUGHT IMPLEMENTS

NEW FEATURES

Hydraulic lift : TRU-TRAC plough with Furrow-width control ; finger-light clutch-brake steering with "hands-off" straight running; slow-running OHV Austin Industrial engine, developing 22 BHP at 1,500 RPM; longer track units, plus scientifically calculated weight distribution, give smooth riding under all conditions at all speeds.

The New Bristol '20' has much more power than its predecessor, which had already established its reputation as the Tractor with the greatest drawbar pull for its size in the world. Yet it is still only 3′ 3″ wide overall, and only 14′ from radiator to back of 2-Furrow Plough.

BRISTOL TRACTORS LTD., Earby, via Colne, Lancs

Fig 2.3 The first Bristol 20 crawler tractors were sold in 1948.

Plate 2.22   The MG2 was the first model of the Ransomes Motor Garden cultivator.

provided to help make sharp turns.

Petrol and paraffin versions of the Bristol 20, both priced at £480, were available in 1948, electric starting was an extra £22 10s 0d and although a power take-off was standard, an additional £47 10s 0d was added to the bill for a rear hydraulic lift and linkage. The Austin 16 engine was replaced with an A70 industrial engine during the latter part of its production run, which came to an end with the introduction of the Bristol 22 in 1952.

The new model was similar to the Bristol 20 except for modifications to the petrol and paraffin engines. The tracks were improved and could be adjusted to give track centre settings from 30 in. to 44 in. with 7 in. wide plates and 33 in. to 47 in. with 10 in. track plates. The basic Bristol 22 tractor cost £742 in 1953 when a 23 hp Perkins P3 diesel engine was added to the list of optional extras. More powerful Bristol crawlers were introduced including the Bristol 25 but many were too big for market garden work.

Approved equipment for Bristol crawlers included rear mounted ploughs, a toolbar with ridging bodies, cultivator tines and hoe blades, power take-off driven circular saw, post-hole digger and winch.

The Ransomes Motor Garden Cultivator was introduced in 1936 following extensive testing of prototypes in collaboration with Roadless Traction who developed the rubber-jointed track for the little blue crawler. Early prototype designs included a tracked machine with the driver walking behind but this was dropped in favour of the more efficient ride-on tractor. Following a good reception at its initial public

Plate 2.23   The MG5 replaced the MG2 in 1948.

Plate 2.24 The 6 hp Sturmey Archer side-valve engine provided the power for this early Ransomes MG 2 to haul this matching Ransomes plough.

demonstration held near Evesham on 29 April 1936, production of the Ransomes MG2 Motor Cultivator started at Ipswich. The MG2 cost £135 and was sold in considerable numbers to smallholders, market gardeners and fruit growers. It also proved to be popular for vineyard work in France.

The MG2 had a 6 hp Sturmey Archer 'T' single cylinder, air-cooled, side-valve engine. The Lucas magneto did not have an impulse coupling and the starting handle dog was on a countershaft linked to the crankshaft by a chain drive. The engine, which had a dry sump, was lubricated by a mechanical oil circulation system with a separate oil tank and pump. The dry sump system was already in use on some of the larger Ransomes motor mowers and the Howard Rotavator. A 4 to 1 reduction gear on the engine output shaft supplied power to the centrifugal clutch at quarter crankshaft speed and the clutch engaged drive to a forward, neutral and reverse gearbox which was not really a gearbox at all.

The transmission unit had two inward-facing crown wheels with a central pinion. Depending on whether the driver wished to move forwards or backwards, the gear lever was used to mesh the pinion with the relevant crown wheel and with the pinion mid-way between the two, the drive was in neutral. Steering was by means of two hand-lever-operated dry-band brakes on the differential shafts and with both

tracks always under power there was no slewing or scraping when the MG changed direction.

The 6 in. wide tracks were adjustable for row-crop work with a choice of 28 in., 31 in. or 34 in. settings. The MG2 was 3 ft 6 in. wide, weighed approximately $10\frac{1}{2}$ cwt and ground pressure was 4 psi. A swinging drawbar and a floating bar used to attach a toolbar, raised and lowered with a hand lever, were standard equipment but the 400 rpm power take-off was extra. The handbook stated that the MG cultivator was designed to do two-horse work at two-horse speed and the operator should not attempt to do more than a two-horse job at 2 mph.

A 1937 issue of the Roadless News published by Roadless Traction informed readers that there were certain jobs which the MG2 could do that appealed to the large-scale farmer and it was noted that one such farmer in the Eastern Counties was using six Ransomes Garden Cultivators almost entirely for inter-row cultivations in his sugar beet. Other farmers living within big centres of population were devoting areas of arable land to growing vegetables and were known to be buying MG tractors for this purpose.

The 'T' type engine was replaced with an

Plate 2.25 The MG6 was introduced at the 1953 Smithfield Show.

Plate 2.26   Raising and lowering the toolbar on Ransomes MG with the hand-lift lever was a good way to develop the biceps.

Plate 2.27   Tasks for the four-wheel drive conversion of the MG6, known as the Ransomes ITW industrial tractor, including hauling trailers and shunting railway trucks.

Plate 2.28   The Ransomes ITC industrial version of the MG motor cultivator.

improved Sturmey Archer 'TB' engine in 1938, an impulse coupling in the Wico magneto improved starting, the cooling fan was belt-driven and the starting-handle dog was on a layshaft linked by a pair of gears to the crankshaft.

A much improved MG5 replaced the MG2 in 1948. It had a 600 cc petrol engine with dry sump lubrication system, fuel lift pump, Wico A type magneto and the starting-handle dog was on the crankshaft. The dry sump lubrication system had two pumps, one supplied oil under pressure into the engine while the second sucked it out and returned it through a filter to the tank. The 4 to 1 reduction gearbox on the engine output shaft and centrifugal clutch, which engaged the drive at approximately 600 rpm, were retained. The MG2 gearbox was used on the new tractor but the gear lever was relocated and the power take-off shaft was increased to 700 rpm. The instruction book again pointed out that the MG5 was designed to do two-horse work at two-

Plate 2.29   Track retainers and mudguards were new features on the MG40, which made its debut in 1960.

horse speed but Ransomes had acknowledged the advance of mechanisation by increasing the top speed from 2 to $2\frac{1}{4}$ mph! The MG2 fuel tank was at one side of the engine and a distinguishing feature of the MG5 was the relocation of the petrol tank under the driving seat.

MG5s were sold in Australia and a hydraulic ram kit for the tool bar was made in that country. The Neville hydraulic lift attachment, designed by a man of that name, was said to provide finger-tip control of all mounted equipment and was available in the UK for £89 15s 0d. The attachment could be fitted by the MG's owner with the only necessary modification being the removal of a small section of the floor plate which could be done in a few minutes with a hacksaw.

The Ransomes MG6 market garden cultivator made its debut at the 1953 Smithfield Show. It retained the basic features of its predecessor but a new three forward and reverse gearbox was a notable improvement. A tvo conversion kit, already an option for the MG5, was available for the new tractor. The centrifugal clutch was retained but the reduction unit on the engine was dropped in favour of the new gearbox, which gave forward speeds of $1\frac{1}{8}$, $2\frac{1}{4}$ and 4 mph. A hand lift toolbar was standard but a new hydraulic lift and power take-off shaft needed to drive the hydraulic pump were optional. The MG6 with a petrol or tvo engine and hand lift cost £305; the price with hydraulic lift, linkage and power take-off was £357.

Two versions of an industrial MG6 motor cultivator were announced in 1956, the Industrial Tractor (Crawler model) or ITC retained the standard tracks but the ITW was a four-wheel drive industrial tractor on pneumatic tyres. Rubber blocks, made by Roadless Traction, could be bolted to the ITC track plates to prevent surface damage when working on roads or concrete. A dumper version of the ITC was also made. It had a rear-mounted engine to make space for a lever-operated self-emptying dump box at the front. The ITW had four pneumatic-tyred wheels coupled in pairs by heavy roller chains to give a rigid four-wheel drive with differential steering similar to modern skid-steer loaders. Ransomes industrial tractors were used to move goods wagons in railway sidings, tow trucks and trailers around factories, also for bulldozing and sweeping with a power take-off driven rotary brush.

An increased range of Ransomes implements for the MG, described elsewhere, was made

Fig 2.4   The fibreglass bonnet and extended track guards became optional fittings for the Ransomes MG40 in 1962.

for MG tractors in the early 1950s including mounted and trailed conventional ploughs, a reversible plough, front and rear mounted tool-bars, disc harrows and a potato lifter. Sprayers and trailers from other manufacturers were approved by Ransomes for use with the MG6.

The reign of the little blue crawlers was coming to an end when the MG40 was announced in 1960. Buyers could choose a petrol, vaporising oil or diesel engine, the centrifugal clutch was retained and the transmission was improved with a three forward and three reverse speed gearbox which transmitted power through a differential and spur gear reduction units to the strengthened rubber-jointed tracks. The MG40 matched the 4 mph top speed of the MG6 but drawbar pull was 1,100 lb in low gear compared with 900 lb for the MG6 and 600 lb for the MG2. Prices ranged from £370 to £495 depending on engine type and implements lift system.

The single-cylinder side-valve 600 cc petrol and tvo air-cooled engines, which developed about 8 hp at 2,100 rpm, had a diaphragm fuel-lift pump, Amal carburettor, Wico magneto and wet sump lubrication with an oil pump and filter. The alternative 8 hp diesel engine was a two-stroke overhead valve air-cooled engine unit with force-feed lubrication. Fuel consumption was approximately 3 pints per hour and the hand-cranked engine was started with the aid of an ignition wick.

Whitlock Bros of Great Yeldham in Essex,

famous for their red farm trailers, made a load-ing shovel and a dump truck based on the diesel engined MG40 in the early 1960s. The Whitlock dumper, unlike the earlier Ransomes ITC, had the manually tipped dump hopper at the back of a standard tractor with the seat rearranged so the driver sat facing the load. The shovel was attached in the same way and the rams were operated by a live hydraulic system with an engine driven pump.

Ransomes gave the MG40 a new look in 1962 by adding a fibreglass bonnet to pro-tect the engine from overhanging fruit tree branches. New steel track guards with optional moulded fibreglass front extensions gave in-creased clearance above the tracks which ran on redesigned load rollers and idler wheel hubs with needle roller bearings. The steel track guards were standard but the fibreglass ex-tensions and bonnet were optional fittings.

Improved tractor design and the demand for more power brought manufacture of Ransomes MG40 tractors to a close in 1966. During its

30 year production run, more than 15,000 MG Motor Cultivators were made at Ipswich, about 3,000 were MG2s, with 5,000 MG5s, a similar number of the MG6 and approximately 2,000 MG40s.

Self-propelled toolbars were another option for market gardeners and smallholders looking for ways to reduce the hand labour required to grow vegetables, potatoes and sugar beet during the 1940s and 1950s. The Bean self-propelled toolbar, David Brown 2D and Wild Midget were real labour savers provided the rows were wide enough apart for the toolbar's pneumatic-tyred wheels.

Humberside Agricultural Products of Brough in East Yorkshire introduced the three-wheel Bean row-crop tractor or self-propelled toolbar in 1946. It had a Ford 8 hp petrol engine and gearbox with a Ford 10 cwt rear axle unit to drive the pneumatic-tyred rear wheels. Electric starting and independent brakes were standard equipment. The driver sat in front of the engine with an unobstructed view of the work and

Plate 2.30   The Bean self-propelled toolbar set up for drilling six rows at a time. Vegetable seeds between the size of a lettuce seed and a pea could be sown and tines on the rear toolbar removed the wheelmarks.

steered the single front-wheeled toolbar with a curved tiller handle. The Bean toolbar was used for hoeing, cultivating, drilling, fertiliser spreading and spraying.

A four-wheeled version of the Bean row-crop tractor, also with tiller steering, priced at £295 was exhibited at the 1950 Smithfield Show, a Bean 5 row hoe was £45 and a five-row drill with large capacity hoppers was £68. A sales leaflet issued at the show pointed out that the three-wheeled tractor, which cost £280, was still available to the discriminating grower wishing to carry out intensive cultivations on a smaller scale. The leaflet also informed prospective buyers that the Bean was so easily handled that the operator would not be tired by tea time and would be able to work overtime without fatigue.

The Wild Midget toolbar, a very basic three-wheeled chassis with a single steel driving wheel at the rear and two front wheels for steering, was exhibited at the 1947 Royal Show. The Midget had a 3 hp J.A.P petrol engine mounted above the rear wheel, which was chain driven through a three forward and reverse gearbox at speeds of $\frac{1}{2}$ to $2\frac{1}{2}$ mph. The driver sat close to the ground on an adjustable hammock seat suspended above the tool frame, which provided an unobstructed view of the rows of plants being hoed, etc.

The Midget had dual steering with a foot-operated rudder bar and a hand lever linked to the steel front wheels. The Midget's underslung tool frame in two 3 ft 6 in. sections was raised and lowered with hand levers and the dual-control steering enabled the driver to lift or drop the tool frame with the hand lever while steering the Midget with a foot on the rudder bar. A second hammock seat could be attached to the frame when it was desirable to have another person riding on the machine for close hand work. It was claimed to be possible to drive the Midget in low gear and hand weed or thin rows of plants while steering the machine with the foot control. The front wheel track was adjustable between 4 ft 1 in. to 5 ft 11 in.

The Allis Chalmers 10 hp Model G self-propelled tool carrier (Colour Plate 3) was made in America between 1948 to 1955 for smallholders and market gardeners. A few examples of this tractor were sold in Great Britain and continental Europe. Power from the rear-mounted four-cylinder Continental petrol engine was transmitted to the wheels through a

Fig 2.5   The Lanz Alldog tool carrier was launched at the 1954 Smithfield Show.

Plate 2.31   The David Brown 2D was basically a self-propelled tool carrier with a rear power take-off and drawbar. It could tow various implements including a trailer, drill and manure spreader.

four forward and reverse gearbox with independent brakes and hydraulic lift as standard equipment.

Levertons of Spalding introduced the diesel-engined Lanz Alldog tool carrier to British growers in 1954. Various implements were made for the Alldog and were attached to a series of holes on each side of the frame. A purpose-built sprayer was made by Allman of Chichester: the 150 gallon tank carried on the tool carrier frame supplied chemical to the power take-off driven pump and 31 ft 6 in. spraybar.

David Brown Tractors chose the 1955 Smithfield Show to launch the 2D self-propelled tool frame. A narrow model for orchards and vineyards was added in 1957. The 14 hp two-cylinder, four-stroke air-cooled diesel engine balanced by an idler piston and cylinder, single-plate dry clutch and four forward speed and

reverse gearbox were mounted above the rear driving axle. Shoe brakes on the transmission half shafts with independent brake pedals and wheel track adjustment from 40 in. to 68 in. completed the basic specification.

The driving seat in front of the engine provided a clear view of the work done with the under-slung toolbar raised and lowered by the David Brown Air-Light lift system. This unusual design used a compressor, driven by the tractor engine, to supply compressed air to twin lift cylinders linked to the tool frame. The central 4 in. diameter tubular chassis served as an air reservoir with

Plate 2.32   A hydraulic motor on the left-hand side of the frame is used to drive the seeder units on this Maskell row-crop tractor.

a tapping-off point for inflating tyres. The mid-mounted tool frame was attached to the 2D by fixing both depth wheels at one end of the bar and wheeling it under the hitch points.

A forward-facing power take-off shaft from the front of the gearbox with its speed related to the 2D's ground speed was standard but a live rear power shaft was an extra. Other optional equipment included a rear-mounted toolbar operated by the air lift, rear belt pulley and electric lighting. Numerous implements, some made by David Brown and others approved for use with the tractor included hoes, cultivators, a down the row thinner, mid-mounted mower and reversible plough.

Just over 2,000 David Brown 2Ds with less than 400 of them in the narrow build had been made by 1961 when production ceased. Some of them are still being used in 1995.

An Enfield 100 twin-cylinder, air-cooled diesel engine provided the power for the Maskell row crop tractor made at Wilstead near Bedford. The engine, clutch, three forward and reverse gearbox and reduction differential connected by chain drives to the wheels were mounted above the twin rear driving wheels. A single, tiller-steered front wheel carried the front end of a hollow rectangular section chassis. A four-wheel version with tiller steering was available at extra cost. The hydraulic lift system and a hydraulic motor, which could be attached at various points on the chassis to drive certain implements, were optional equipment. Barfords of Belton, a member of the Aveling Barford Group, produced the renamed Barford-Maskell row crop tractor in the early 1960s.

There was still a demand for small self-propelled tool carriers in the mid 1970s and this was met by Russells of Kirbymoorside with the 20 hp 3D powered tool carrier. It had a rear-mounted, two-cylinder, air-cooled diesel engine, hydrostatic transmission with a hydraulic motor in each rear wheel, independent disc brakes and hydraulic lift. A single lever controlled the infinitely variable speed to a maximum of 9 mph. Another lever raised and lowered the

Plate 2.33 The Evenproducts Growmobile motorised tool carrier could be used to drill seeds, place fertiliser and apply chemical granules at the same time.

mid-mounted tool frame and a third operated the track eradicating tines behind the rear wheels. Cultivator tines, hoe blades and seeder units were among the attachments for the 3D tool carrier, which cost approximately £4,500 in 1979.

Evenproducts of Evesham, well known for their irrigation equipment, introduced the Grow-mobile multi-purpose growing machine in the mid 1980s. There were two models of this self-propelled tool frame with a rear-mounted air-cooled Honda petrol engine and hydrostatic motors in both rear wheels. The smaller 8 hp Mark 1-R, which cost £4,675 in 1987, had a ground clearance of 26 in. and the 11 hp Mark 2-R Growmobile with 33 in. clearance cost £5,225.

A heavy-duty alternator provided the power for the electric lift mechanism, which raised and lowered an underslung toolbar used for drilling, cultivating, hoeing, etc. Front- and rear-mounted equipment including a fertiliser spreader, granule applicator, sprayer and weed wiper were driven by a 12 volt electric motor, which made it possible to use front, underslung and rear-mounted equipment at the same time.

3 Hoes, Ploughs and Cultivators

HOES

A wide choice of single- and two-wheeled push hoes was available to help reduce the hand labour needed to remove weeds growing between rows of plants in domestic and market gardens during the 1930s and 1940s. Small self-propelled motor hoes, popular in the late 1940s and early 1950s, took still more drudgery from the never ending battle with weeds. This is also the period when two- or four-wheel garden tractors with hoe blades on a toolbar were replacing the horse hoe on many market gardens and smallholdings.

Hoeing between rows of plants with a Murwood, Jalo, Planet, Wrigley or other make of single two-wheel push hoe was a familiar sight in the late 1940s. Single-wheel hoes were pushed along between the rows with a pair of L hoe blades set as close as possible to the plants. Two-wheeled hoes were used with the wheels either straddling a single row or with both of them running between two rows.

Murwood Agricultural made M & G single- and twin-wheeled push hoes, which were also used with cultivator tines, a rake, a plough and a small ridging body. The M & G sales leaflet of 1950 included the single row M & G push hoe with four cultivator tines and two 8 in. hoe blades at £3 carriage paid, a seeder unit cost £2 5s 0d and a kit to convert the hoe to a wheelbarrow (Fig 4.4) was £2. The twin wheel M & G hoe with hoe blades, cultivator tines and leaf guards cost £5.

The American Planet Junior single and two-wheeled, wooden-handled push hoes were sold in their thousands throughout the world. There were several models including the No. 4 with one wheel which cost £4 19s 9d in 1940 and the No. 25 with two wheels at £5 18s 9d. Hoe blades, cultivator tines, a small plough and a seeder unit

were included in the price. A special bracket and extra wheel could be obtained to convert the No. 4 to a two-wheeled hoe.

J.T. Lowe at Wimborne in Dorset made Jalo single- and two-wheeled push hoes and Jiffy seeder units in the late 1940s. Publicity material in 1950 described the new hoe as the triumphant result of countless experiments with varying materials under actual working conditions that at last made it possible to get rid of the back-breaking work of hand hoeing. Potential buyers were also informed that the streamlined shape of the unit gave perfect control and even the most

Plate 3.1 The M & G two-wheel hoe had leaf guards to deflect the leaves away from the wheels.

troublesome weeds were severed instantly!

Supplied direct from the manufacturers, the Jalo hoe complete with a pair of hoe blades cost £3 7s 0d carriage paid. Attachments for cultivating, drilling, raking and ploughing in cultivated ground were available at extra cost.

J.T. Lowe adopted Jalo as the company name in 1951 and push hoes were made well into the 1960s. The Jalo Gardener with a 'U' shaped tubular handle, widely advertised during the late 1950s, was claimed to hoe up to twelve times faster than was possible by hand. Complete with a pair of hoes and small plough, the Jalo Gardener cost 5 guineas in 1959. Hire purchase terms were available with a down payment of £1 6s 0d and four more instalments of the same sum. It could be used to ridge, drill, cultivate, rake and the twin-wheel model could be converted to a wheelbarrow or fitted with a Sheen Flame-Gun to burn off weeds.

Fig 3.1   A full set of tools was included in the price of single- and two-wheeled Planet Junior push hoes.

SAY 'GOODBYE' TO WEEDS
—THE MODERN WAY!

JALO Hoes are designed for every GARDEN, AL-LOTMENT, SMALLHOLD-ING, NURSERY & FARM and are the result of count-less experiments under actual working conditions. Not until you have actually tried the JALO will you really believe how much *Easier, Quicker* you can remove weeds. Whether hoeing, cultivating, raking or ploughing with your JALO Hoe, you'll be de-lighted with its perfect performance.
PRICES :   Single-wheel Model, 70/-; Twin-wheel Model, 105/-. Jalo Hoes are only obtainable direct from the Makers.

Fig 3.2   The Jalo single-wheel push hoe, introduced in 1950, cost £3 3s 0d but within a year the price was increased to £3 10s 0d.

Tractors (London) Ltd and Jalo combined their resources to produce the Jalo Trusty motor hoe in the early 1950s. It had a ⅓ hp engine with a chain drive to friction rollers held against the hoe wheels by the weight of the engine. Drive was engaged with a centrifugal clutch and a lever was provided to lift the engine unit off the wheels so that the hoe could be pushed into confined spaces. Publicity material at the time advised there would be no more struggling or straining with out-of-date hand hoes. No more grunts and groans from back-breaking hand-work as the Jalo Trusty hoe would easily and quickly remove weeds ten times faster than before.

There was plenty of competition for push hoe sales in the 1950s when manufacturers included

Plate 3.2   This early 1950s Jalo Trusty motor hoe cost £29 10s 0d.

Fig 3.3   The Pulvo hoe was said to pound the earth to a fine tilth.

George Bignell, Industrial and Agricultural Improvements, Landmaster, Pulvo and Wessex Industries. The single-wheel New Colonial cultivator was made by Geo Bignell at Sutton Coldfield in the early 1950s. The Wrigley push hoe manufactured by Wessex Industries at Poole could be used to hoe, rake, plough and cultivate. The single-wheel Wrigley cost £4 5s 0d in 1959 and the two-wheel model was an extra £2. An advertisement described the Wrigley hoe to be undoubtedly the most practical implement obtainable at such a remarkably low price.

Publicity material in 1951 stated that the Ro-Lo cultivator bridged the gap between man-power and horsepower and offered a new era of ease and prosperity to all engaged in working the soil for profit or pleasure. Made by Industrial and Agricultural Developments at Great Malvern, the Ro-Lo was carried on a smooth roller instead of wheels with a short toolbar behind it for hoe blades and other accessories. The Ro-Lo cultivator cost £3 10s 0d. Cultivator tines and a ridging body were extra.

The Pulvo garden hoe ran on an open roller with a frame for hoe blades. The vibratory action of the off-set bars on the roller were said to pound the earth to a fine tilth faster and more easily than would be thought possible. The Pulvo appeared in the late 1950s and cost 5 guineas with the usual set of attachments.

The push hoe made life easier for gardeners and small-scale growers but a small motor hoe was an even easier and more efficient way of killing weeds in row crops. Some growers used a garden cultivator with a front or rear tool-bar to remove weeds, others preferred a rotary cultivator but these machines were inclined to throw soil over the small plants. Hoe blades were among the toolbar attachments for many garden tractors but small motor hoes such as the British Anzani, Colwood Hornet, Ransomes Vibro-Hoe, Commando, Coleby Minor, Teagle Digo and Pegson Unitractor were more suited to small-scale producers during the late 1940s and early 1950s.

The British Anzani Motor hoe cost £39 10s 0d complete with hoe blades and grubbing tines when it was introduced in the late 1940s. The 1 hp Anzani-J.A.P four-stroke engine ran all day on 5 pints of petrol and forward speeds between 1 and $2\frac{1}{2}$ mph were set with the throttle lever. It had a centrifugal clutch with reduction gears and independent wheel clutches. The overall width of the hoe was 8 in. but the wheel track could be increased to 13 in. by adding spacers to the axles when hoeing wide rows. A sales leaflet pointed out that the Anzani Motor Hoe was

Fig 3.4   The Wrigley push hoe cost £2 19s 0d in 1954; five years later it was £4 5s 0d.

not a tractor and was intended for work on properly prepared soils. The simplicity of control and perfect balance were said to give expert results after very little practice and make work a pleasure.

Dashwood Engineering introduced the single-wheeled Colwood Model A and Model B motor hoes in the mid 1940s. The Model A with a front bumper bar had a 1.2 hp J.A.P 2A engine to drive the single steel wheel through a three-speed gearbox. The Model RA rotary hoe was similar to the motor hoe but had a 1.2 hp Villiers four-stroke engine with a $9\frac{1}{2}$ in. wide detachable rotor.

The Model B Colwood motor hoe had a two-speed Dashwood–Albion gearbox and a J.A.P 2A engine but within five years it was replaced with a governor controlled 1.2 hp

Plate 3.3   Planet Junior push hoe tools could be fitted to the British Anzani motor hoe tool frame.

Plate 3.4 The Clifford Model A1 garden tractor with front-mounted hoes.

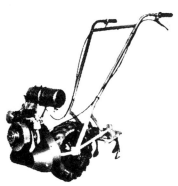

It's THOROUGH CULTIVATION *plus*

EASIER WORKING

—solo wheel makes them simple to steer.

NO HEADLAND REQUIRED

Colwood implements can be turned on their own axis.

MORE POWER

—let the machine do the work.

PERFECT BALANCE

—essential on rough ground and slopes.

FULL RANGE OF ATTACHMENTS

—for hoeing, cultivating, cutting long grass, mowing, spraying, pumping, hedge trimming, carting, etc.

with

Colwood

HORNET MOTOR HOE

Particularly suitable for light hoeing. Supplied with "planet" type too. *Specification: Engine,* Villiers "Midget" engine Mk. 11, 2-stroke 98 c.c. *Controls,* Clutch lever. Twist grip throttle.

GARDEN TRACTOR

Model B. The popular, well-known garden tractor for inter-row hoeing and cultivating. Overall width 13 in. and only 9 in. wide below the frame without guards. *Specification: Engine,* 1.2 h.p. governor-controlled 4-stroke Villiers Mk. 10. *Starting,* Lever. *Gearbox,* Dashwood-Albion. 2 forward speeds.

HORTICULTURAL AND AGRICULTURAL IMPLEMENTS

Plate 3.5 Colwood Hornet and Colwood Model B motor hoes.

Villiers Mark 10 four-stroke power unit. Both engines were started with a hand lever. A full range of attachments including cultivator tines, grass mower, hedge trimmer, pump and trailer, was made for the Colwood Model B.

The Colwood Hornet motor hoe with a Villiers Midget Mark II 98 cc two-stroke engine appeared in 1951. The gear-driven land wheel was positioned centrally between the engine and the $9\frac{1}{2}$ in. wide tool frame. As well as the range of accessories made by Dashwood Engineering, Planet Junior push hoe tools could be used on the Hornet's tool frame.

The Coleby Minor motor hoe with a J.A.P 2A engine and centrifugal clutch to engage drive to the wheels was less than 8 in. wide. Top speed was $1\frac{1}{2}$ mph and ratchets provided independent drive to the wheels. Small garden tractors such as the Pegson Unitractor and the American designed Gravely were also used in the late 1940s for hoeing in row-crops.

The single-wheel Gravely motor cultivator with hoe blades on a tool frame in front of the engine could be used to hoe row-crops at speeds

Plate 3.6 A sales leaflet suggested that the Coleby Minor motor hoe could, with very little tuition, be handled very easily by either sex.

Plate 3.7   The Gravely motor cultivator with a front hoe.

Plate 3.8   The Trusty tractor with hoes.

of 3 mph. The Pegson Unitractor, similar to the Eaglesfield Unitractor made in Indianapolis, had a front toolbar with a Villiers engine mounted inside its steel rimmed wheel.

The Teagle Digo motor hoe made by W.T. Teagle at Truro in Cornwall during the late 1950s had a 49 cc Teagle two-stroke engine with a recoil starter mounted on the handlebars. A very basic drive arrangement propelled a pair of steel wheels similar in appearance to the spinning disc on a fertiliser broadcaster. Hoe blades and other cultivating tools were bolted to a narrow tool frame behind the wheels. Other fittings for the Digo, which cost £45 in 1958, included a lawn mower and a flexible drive shaft for a hedge trimmer.

Plate 3.9   J.C. Rotor Hoe in use on the Barford Atom.

An unusual hoe attachment suitable for one- and two-wheeled garden tractors was made in the mid 1950s by A.M. Russell at Edinburgh. The J.C. Rotor Hoe consisted of a horizontal weed cutting blade below and behind a ground-contact-driven rotor similar to a lawn mower cylinder with seven inclined blades. The Rotor Hoe was made in 8, 14 and 20 in. widths and all three models could be extended by 4 in. to hoe 12, 18 and 24 in. rows. The smallest Rotor Hoe with three cutter blades cost £9 15s 0d in 1953. Sales information suggested the Rotor Hoe was particularly efficient in reasonably dry soil conditions and it made weeding by normal hand hoeing obsolete.

Described as the only powered draw hoe capable of working in any soil, the Commando power-driven hoe, which cost £65, was made by Power Hoes Ltd of London SW3 in the early 1950s. The wheels and draw hoe blades were driven by a 1 hp four-stroke engine which used a pint of petrol per hour. The throttle and clutch levers to engage drive to the wheels and hoe blade crank mechanism were on adjustable handlebars, which could be offset to nine different positions on the 9 in. wide Commando hoe. Hoeing width was instantly adjustable from the handlebars and depending on size of blade it was possible to hoe between rows from 9 in. and 22 in. apart in a single pass.

The Commando was said to go anywhere and tackle anything. It could work against a wall or

hedge, incorporate mulches or fertiliser, earth up plants or scrape soil away from the base of onions and similar crops.

The self-propelled Ransomes Vibro Hoe was another mid 1950s mechanical hoe, the reciprocating blades moved backwards and forwards through a small arc to cut through the weeds and any remaining soil was shaken from the roots by short lengths of bicycle chain secured to the back of the blades. Although prototypes had two wheels, the Vibro Hoe had a single wheel to keep it as narrow as possible. The hoe blades were 6 in. wide but rows up to 18 in. apart could be hoed with extension bars on the blades. The Vibro Hoe's vibrating motion was also an advantage when using cultivator tines.

A Villiers 98 cc two-stroke engine provided the power and a centrifugal clutch engaged the drive through a roller chain and gears to the land

Plate 3.10   Hoeing five times faster than it was possible with a hand hoe was a good reason for buying the Commando power-driven hoe.

Plate 3.11   According to the sales literature, the Ransomes Vibro Hoe would banish once and for all the tiresome, backaching drudgery associated with older methods of hoeing.

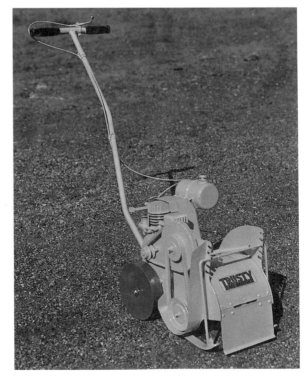

Plate 3.12   The Trusty Whirlwind hoe and grass cutter.

Plate 3.13   The Trusty Weedsweeper.

Plate 3.14   This Bean row-crop tractor with a five-row hoe cost £340 in 1951.

wheel. The hoe blades were driven by a system of cranks and connecting rods arranged so that when one blade was on its forward stroke the other was moving backwards. The number of hoeing strokes per minute was directly proportional to the Vibro Hoe's forward speed. A dog clutch disengaged the drive to the land wheel and the hoe mechanism, this was used before starting the engine as it was necessary to set the throttle at maximum speed. The dog clutch had a tendency to seize and when it was dismantled the springs and ball bearings were likely to fly in all directions. Replacement parts for the dog clutch were in constant demand from Ransomes spares department!

The Trusty Whirlwind combined rotary hoe and grass cutter with a $\frac{1}{3}$ hp engine was introduced by Tractors (London) Ltd in 1957. Transmission was by vee-belt to a rotor disc with a set of pins which removed the weeds. The Trusty Weedsweeper, also introduced in 1957, had the same engine with vee-belt drive to a 6 in. wide rotary cultivator.

There was a great deal of walking to do with push hoes and cultivators and it was not long before mounted steerage hoes for the Ferguson 20, David Brown 2D, Bean, Garner and other tractors brought relief to tired legs.

Flame guns provided an alternative method of killing unwanted vegetation and they had much in their favour when dealing with perennial weeds. Several companies including Bering Engineering of Camberley, McAllan Appliances of London and Sheen of Nottingham were selling flame guns in the 1950s. Sheen introduced their flame guns early in the 1950s and within ten years they were making 14 different models and some were still being manufactured in 1975. Sheen flame guns ranged from a small hand-held torch to large flame guns for use with small garden tractors or push hoes. A 1962 Sheen advertisement admitted that a flame gun would not kill perennial weeds in one operation but explained that their two-stage system gave fully effective weed control in a fraction of the time taken by conventional methods.

The Sheen X300, a typical hand-held flame gun, had a 1 gallon fuel tank, a hand-operated pump to pressurise the container, a pressure gauge, control valve and burner which produced a flame of approximately 2,000°F. A hood was used to restrict the spread of flame in confined spaces.

Plate 3.15   Hoeing with the Ransomes MG and front-mounted toolbar.

A wheeled version of the X300 gun had a hinged hood and was recommended for inter-row weeding and for burning off weeds on paths and crazy paving. The Sheen Flamewand, looking rather like an overgrown bicycle pump with a tank holding enough fuel for about 30 minutes, produced a torch-like flame suitable for weeding in very confined spaces.

PLOUGHS AND CULTIVATORS

The spade and fork have been basic tools for working the soil for centuries and even these simple implements were the subject for more than one inventive mind in the late 1940s. The revolutionary Terrex semi-automatic garden spade, first made in France, was and still is used by applying foot pressure to a spring-loaded lever at the base of the spade handle until

the blade is a full spit deep. A tug on the handle released the spring, which flicked the blade forward, loosening and partly lifting the soil, which is then turned in the normal way. The spade was advertised for £3 19s 6d post free in 1951, it cost £7.95 in 1973 and it can still be purchased in 1995 with a tenfold price increase at £79.95.

Terrex also made a semi-automatic garden fork in the 1950s but there was a cheaper way to take some of the backache from digging. The Easi-Digger, advertised in 1952, was a device consisting of a lever and a powerful spring attached to a spade or fork handle. It cost £1 9s 9d and was claimed to take the hard work out of digging.

Cultivator tines and ploughs made for push hoes were quite effective on prepared soil but engine power made life easier when ploughing hard ground. Miniature ploughs and cultivator tines were often included in the price of motor hoes and some of these machines could be used with tools made for Planet Junior push hoes but here again they were more effective in loose soil. Ample power for breaking undisturbed soil with a plough or cultivator was provided by various makes and models of two-wheel garden tractors, ride-on tractors and small tracklayers.

Toolbars for tined implements were made for some two-wheeled garden cultivators but many of these machines were only used for rotary cultivating. The more powerful two-wheeled tractors such as the British Anzani, B.M.B Plow-Mate and Trusty, considered an alternative to a horse, were used for ploughing and cultivation work. It was hard and exhausting work to walk day after day behind a two-wheeled tractor with a single-furrow plough and the chance to ride on a B.M.B President, Byron, Garner, Gunsmith, OTA and other three- or four-wheeled tractors was less tiring and certainly more productive. Pricewise these tractors were not very competitive compared with the Ferguson TE20 and its refined hydraulic system. For this reason smallholders and market gardeners were able to justify the extra cost of a little grey Fergie which brought new found simplicity to ploughing, cultivating, discing, rolling and other fieldwork.

Ransomes, Sims & Jefferies of Ipswich made miniature versions of their ploughs and

Fig 3.5   A selection of Sheen flame guns made in the 1960s.

Fig 3.6   The Bering flame gun.

cultivation equipment for MG crawler tractors. The TS25 two-furrow trailed plough for the MG2 could be converted to a single-furrow model and when used in that form could plough between 8 and 10 in. deep depending on conditions. The TS30 single furrow and TS31 two-furrow hand-lift ploughs came next and both had a small screw handle on the rear wheel to adjust ploughing depth and keep the plough level from front to back.

The self lift TS42 single-furrow plough with screw handles within easy reach of the driving seat for adjusting depth and levelness were a great improvement. The TS42A, with much lower screw handles for ploughing under tree branches in fruit orchards, and a strengthened TS42B were added at a later stage. The TS65 single-furrow mounted plough was made for the hydraulic linkage on the MG6 and the MG40, which replaced it in 1960. The TS65 was a right-handed plough and when Ransomes added the left-handed TS66 to their range, it was possible to reversible plough with the MG by hitching the two ploughs side by side on the hydraulic linkage.

Compact tractors from Japan were popular in Great Britain by the late 1970s and a range of ploughs and cultivators designed for the new generation of light tractors was made by the Japanese and some British implement

(*text continues on page 84*)

Fig 3.7   Terrex semi-automatic spade has been made for 50 years.

Plate 3.16   Hoe blades, weeder tines and a ridging body were among the attachments for the Merry Tiller toolbar.

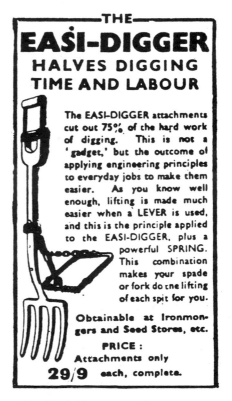

Fig 3.8   The Easi-Digger was claimed to take the hard work out of digging.

THE VERSATILE *MAYFIELD*

Ploughing conserves the soil.

Why not conserve your energy with a Mayfield, and in addition save both time and money. Besides ploughing it will tackle 14 other jobs, all attachments being fixed easily and quickly. 3 models are available, all of which have 4-stroke engines and 3-speed gearboxes, and are of robust construction, yet light to handle, suitable for lady operators. Reverse gear can be supplied as an extra.

The price is low, too, and is the best value for money on the market today.

MAYFIELD ENGINEERING (Croydon) LIMITED
Church Street, Dorking, Surrey

Fig 3.9   An advertisement for the 1959 Mayfield tractor and plough.

Fig 3.10   Troy Tractivator Mark II with single furrow plough.

Plate 3.17   Late 1940s sales literature claimed that the British Anzani Iron Horse was able to plough at least as much land as two horses.

Plate 3.18   The screw handle on the plough for the B.M.B Plow-Mate was used to adjust furrow width.

Plate 3.19   Ploughing with the Garner light tractor.

manufacturers. Reversible ploughing was an almost universal practice in British agriculture by this time and the availability of reversible ploughs for compact tractors enabled market gardeners and smallholders to follow this trend.

Some market gardeners harvested their potatoes with a mouldboard plough rather than endure the backaching work of lifting the crop with a wide-tined potato fork. Both methods left many potatoes buried in the soil but the potato plough with prongs behind a small ridger-type body left more of the crop on the surface for the hand pickers. Potato ploughs were made for several garden tractors including the Iron Horse, Clifford and Ransomes MG crawler.

Most farmers were using potato spinners or elevator diggers to harvest their potato crops in the 1940s and 1950s. Clifford Aero and Auto made a potato spinner for their rotary cultivators, which made picking crops grown on allotments and market gardens a little easier than it was behind a potato plough. A wide share loosened the ridge and the spinner rotor,

(*text continues on page 87*)

Plate 3.20   The Ransomes TS25 for the MG crawler could work to a depth of 6 in. and when converted to a single-furrow plough the maximum depth was increased to 10 in. The two-furrow model cost £12 nett cash, the single furrow model was £9 9s 0d and skim coulters were 10s 6d each.

Plate 3.21 The TS65 single-furrow plough was made for the MG hydraulic lift.

Plate 3.22 Reversible ploughing with Ransomes single-furrow TS65 and TS66 ploughs mounted side by side on the MG40 hydraulic linkage.

Plate 3.23 Reversible ploughing in the mid 1970s with a Kubota B6000 tractor.

Plate 3.24   An Iseki compact tractor with a Wessex two-furrow reversible plough in 1989.

Plate 3.25   Backache from lifting potatoes with a hand fork could be avoided by harvesting the crop with a potato plough. This model was clamped to the Trusty toolbar.

driven by the rotary cultivator power shaft, moved soil and potatoes sideways on to fresh ground for the hand-pickers.

A mechanical spading machine called the Rotaspa was developed in Holland by Vicon in the early 1960s. The horticultural version had three sets of three spades with a working width of 42 in. and a maximum digging depth of 12 in. A larger 7 ft spading machine was made for the agricultural market. The power take-off driven blades rotated around a horizontal rotor shaft and an arrangement of gears inside this shaft caused each blade to cut, lift and invert a block of soil before returning it to the ground. Sales literature explained that the soil was left spaded in exactly the same way as hand digging and because the blades turned quite slowly, soil smear and panning were prevented. The 42 in. Rotaspa required at least 13 hp at the power take-off and could dig about $\frac{3}{8}$ acre per hour at an average speed of 1 mph.

Self-propelled rotary cultivators (described in Chapter 1) have been used by gardeners and commercial growers since the 1940s to break up soil, prepare seedbeds and cut up weeds and crop residues. Some of these machines, notably those made by Rotary Hoes, were single-purpose rotary cultivators but most garden cultivators had a rotary cultivator, which could be interchanged with a plough and a toolbar for row-crop work.

A new generation of three-point linkage
(*text continues on page 92*)

Plate 3.26 The Clifford potato spinner cost £22 10s 0d in 1950. The large-diameter steel wheels were recommended and cost an extra £11 15s 0d.

Plate 3.27　The three spade Vicon Rotaspa, introduced in the early 1960s, was suitable for market gardens and glasshouses.

Plate 3.28　Howard Bantam Rotavator.

Plate 3.29   Inter-row cultivating with the Howard 400 Rotavator in the late 1960s with the handlebars off-set to avoid walking on the freshly tilled soil.

Plate 3.30   A Kronevator rotary cultivator and Kubota L175 at work in 1975.

Plate 3.31   The 3ft wide Mansley Rotary cultivator, approved for use with the Ransomes MG40 crawler, could cultivate an acre in an hour and a half.

Plate 3.32   When cultivating, the Barford Atom's $1\frac{1}{4}$ hp four-stroke engine used less than $\frac{3}{4}$ pint of petrol per hour.

Plate 3.33   A late 1980s Massey Ferguson 1010 and Dowdeswell Powervator.

Plate 3.34   The Ransomes MG2, with seven cultivator tines on a Ransomes hand-lift toolbar.

Plate 3.35   Disc harrowing with the Ransomes MG6 and a 3 ft wide HR4 harrow.

rotary cultivators for compact tractors appeared in the mid 1970s. They were similar to those made by Rotary Hoes twenty years earlier for the Ferguson 20 and other tractors in the 20 to 30 hp range. Many of these rotary cultivators were off-set on the tractor to remove one of the marks made by the rear wheel and were used round and round an area of land to avoid wheelings on the cultivated surface.

Spike-tooth harrows, cultivators and other tined implements date back to the age of the horse and the same basic tillage implements were made in various sizes for use with hand-pushed tools, two- and four-wheeled horticultural tractors and market garden crawlers. A simple tool frame with a pair of wheels was either included in the price or an optional extra for most two-wheeled garden tractors made in the 1940s and 1950s. The range of tools varied

Plate 3.36   Cambridge and flat rolls, both 4 ft 6 in. wide, were made for Clifford cultivators. The 6 cwt Cambridge roll cost £38 in 1950 and the lighter flat roll was £30.

Plate 3.37   Massey Ferguson 1030 with spring-loaded rigid tine cultivator.

from a few tines and hoe blades to lengthy lists of tillage attachments for the British Anzani Iron Horse, B.M.B Plow-Mate and the Trusty.

An extensive range of cultivation equipment including tines and hoe blades was made for the front- and rear-mounted toolbars used with Ransomes MG crawler tractors. The C29 and C71 were raised and lowered manually with the hand-lift lever on the toolbar. The front-mounted C70 and the C67 toolbars designed for the optional rear-mounted linkage on the MG6 and MG40 were raised and lowered hydraulically and were unlikely to leave the driver with an aching back at the end of a long day. A 3 ft wide set of Ransomes HR 4 trailed disc harrows with 16 in. diameter discs angled from the tractor seat were also made at Ipswich for MG crawlers.

Tillage equipment approved for use with MG crawlers included land rolls manufactured by Hunts of Earls Colne and Gibbs of Bedfont in Middlesex, a steerage hoe made by G.J. Garrett & Sons at Dartford and the Mansley rotary cultivator from Gregson & Monk at Preston.

4 Sowing, Spreading and Spraying

Mechanical aids for sowing seed and spreading fertiliser were being used on some market gardens and smallholdings in the 1940s but the home gardener did this work by hand. A variety of spraying and dusting machines for the distribution of liquid and powdered chemicals was also available at this time but designers appear to have taken little heed of the health risks involved for those using this equipment.

Small hand-held gadgets were made to help the home gardener sow seed sparingly and it was still possible to buy a new seed fiddle in the 1950s. Many market gardeners used a hand-pushed seeder unit in the late 1940s to drill anything from lettuce to peas in continuous rows.

The brush feed drill mechanism, which dates back to the early 1900s, was still in common use in the immediate post-war period. The Feedex brush feed drill had a land wheel driven rotary brush on a shaft in the bottom of the seed hopper and a simple coulter to make a shallow furrow. The brush swept seed through an adjustable-sized hole in a disc near the base of the hopper into a short tube connected to a coulter. The metal seed disc with various diameter holes was rotated to give the required aperture size for many types of seed. The disc also provided a means of varying the seed rate and the hole was closed at the end of a row to cut off the flow of seed.

The Coleby seed drill for two-wheeled tractors

IT'S SOW EASY

Make every crop a competition winner by sowing with a " Shakit."

- Even and rapid sowing.
- Adjustable for different sized seeds.
- Nothing to wear out.
- Reduces need for thinning, thus promoting sturdier growth.
- Indispensable to the serious gardener.

2'6 PLUS 3ᵈ POSTAGE

W. E. BEVAN & CO.,

Fig 4.1   The Shakit seed sower was indispensable to the serious gardener.

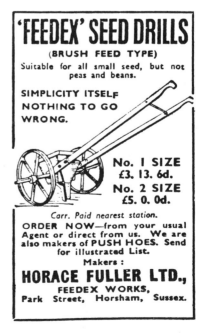

Fig 4.2   The Feedex seed drill was made in the early 1950s.

94

Plate 4.1 The M & G seed drill, made in the 1950s, was suitable for all types of seed up to the size of a pea. The seed hopper held 7 pints and it cost £7 5s 0d.

was an improvement on the rather inaccurate brush-feed mechanism. Patented sweep discs on a wheel-driven shaft in the hopper were claimed to give a more positive delivery of seed than a brush. Seed spacing depended on the number of discs from the shaft and the mechanism was used to place seeds at 10, 15 or even 30 in. apart to save a considerable amount of hand work when the crop was singled.

The American Planet Junior drill, an attachment for the Planet push hoe, was used to sow seed in continuous rows or in small groups. A set of star-shaped cams was provided with each seeder unit; seed rate and spacing were altered by changing the cam. The land wheel driven cam opened a shutter in the bottom of the hopper to release seed into a shallow furrow in groups at intervals of 4, 6, 8 or 12 in. apart. The seed release shutter was held open when drilling seed in a continuous row.

Plate 4.2 The Coleby seed drill could be used with many models of garden tractor made in the 1950s.

Combined Hand Seed Drill and Cultivator

Are popular seeders in all parts of the World. They are easy to push and sow evenly. The opening plow opens a neat, clean narrow furrow and has wide adjustment for depth. Coverers fill in soil evenly. Fitted with marker rod. Provides for planting various seeds by a variable opening with accurate screw adjustment.

The No. 3 will sow practically all vegetable (and some flower) seeds in drills. Continuous or "set distance" drills. A favourite amongst Seed Growers, Market Gardeners, etc.

No. 3. Hopper holds 3 qrts. **£4 19 9**
No. 5. Hopper holds 5 qrts. **£5 11 9**
Carriage Paid England and Wales.

Fig 4.3   *The Planet Junior seed drill. A spare star-shaped cam for altering seed spacing is attached to the handlebars.*

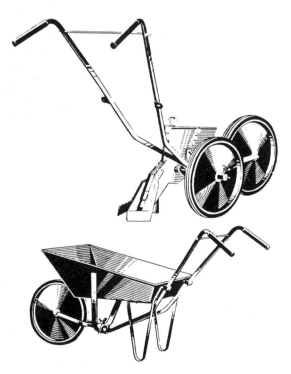

Fig 4.4   *The M & G drill unit for the Murwood push hoe cost £2 5s 0d in the early 1950s. A barrow body costing £2 could also be fitted to the hoe and the enterprising gardener could push the drill unit and seed to the allotment in the barrow and then convert it to a seeder unit!*

A later version of the Planet seeder, used either as a single row, hand-pushed drill or as a multi-row drill with a number of units bolted to a tractor toolbar, could plant an even wider range of seeds. The land wheel-driven seeding mechanism consisted of a wavy agitator disc which directed seed through a round hole in a circular seed plate attached to the bottom of the seed hopper. Planet seeder units were supplied with four seed plates and each plate had a series of different sized holes and one set of seed plates was sufficient to plant any size of seed from clover to beans. The required seed plate was clipped in position under the hopper and rotated to align the chosen seed hole with the outlet point above the coulter.

The seed hopper on the M & G drill unit made for the Murwood push hoe held 1 pint of seeds and the agitator feed mechanism could sow any seed up to the size of a pea. The unit complete with its own drive wheel was attached to the Murwood push hoe frame and secured with a thumb screw in a matter of seconds. A Perspex panel in the hopper enabled the user to check seed flow to the coulter.

Humberside Agricultural Products, manufacturers of the Bean self-propelled toolbar, made a large number of Bean seeder units in the 1950s. The Bean seeder was a complete unit with a drive wheel, rear press wheel, seed hopper and

Plate 4.3   Single-row Bean hand drills cost £9 17s 6d in 1952. The price included the extra seed disc clipped to the handlebars.

seeding mechanism similar to that used for the later type of Planet seeder. Single-row hand-pushed drills for market gardeners and multi-row drills on mid- or rear-mounted toolbars were used to plant larger areas of vegetables and other crops.

The hand-pushed drill had a 4 quart capacity hopper, a shut off lever was provided on the handlebars and a marker scratched a line as a guide for the next row. Pints and quarts were often quoted on 1950s sales leaflets as units of measurement for the hopper capacity of seeder units. Pea and bean seeds were also sold in pint and quart packets.

The Cuthbertson unit drill made at Biggar in Scotland and the French designed SM frame drill for cold frames and glasshouses were also used by market gardeners. The Cuthbertson drill, made in the late 1950s, had a small land wheel driven auger revolving in a slotted bush in the bottom of the seed hopper. A rubber agitator disc above the auger ensured a constant flow of seeds was carried from the hopper in the spaces between the auger threads and taken to the outlet point where they fell into a shallow furrow. Different sized sets of auger and slotted bush supplied with the seeder unit were used to sow a variety of seeds at different rates. The single-row hand-drill cost £33 and individual units for tractor toolbars were £30.

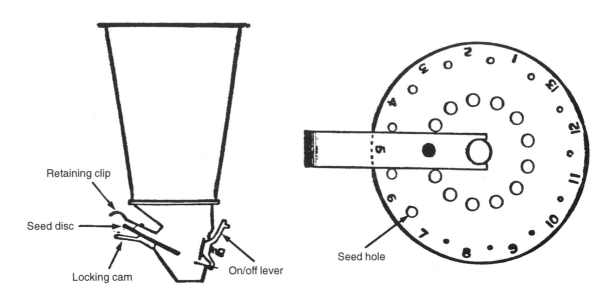

Retaining clip

Seed disc

Locking cam

On/off lever

Seed hole

Fig 4.5   A rotary agitator in the Bean seeder unit hopper forced seed through the selected hole in the seed disc.

Plate 4.4   Drilling carrots with Bean seeder units on the Bean self-propelled toolbar.

Plate 4.5 The British Anzani Iron Horse with three Bean units.

The SM frame seed drill, made from aluminium, consisted of a bank of four small diameter feed rotors driven by a square shaft from the land wheels. The SM drill could be used in a cold frame to sow four closely spaced rows at a time and different feed rotors were used when changing seed type or altering the spacing of the seed in the row.

Precision seeders for vegetable crops and sugar beet were developed by several agricultural machinery manufacturers in the early 1950s. The Cuthbertson and SM seeder units were early examples of spacing drills but the Stanhay belt feed and the cell wheel mechanism used by other makers including Webb and Russell have withstood the test of time.

Stanhay of Ashford in Kent introduced a belt feed precision seeder unit in 1953. Holes punched at equal intervals in a narrow rubber belt carried seed from the hopper to an outlet point above the coulter. The cell wheel mechanism used a similar principle but with this design the seed is carried from the hopper to the outlet point in evenly spaced holes around the outer rim of the feed wheel.

Single-row push drills with belt and cell wheel feed mechanisms became available to small-scale vegetable growers. Market gardeners with larger areas to plant were able to use two or three seeder units on a two-wheeled garden tractor toolbar (Colour Plate 19).

Russell's Exel cell wheel hand seeder cost £80 in the late 1970s. It had a 2 pint capacity hopper, and the aluminium cell wheel was driven by its rubber-tyred land wheel. Russell's sales literature indicated that seed could be spaced at 1 or

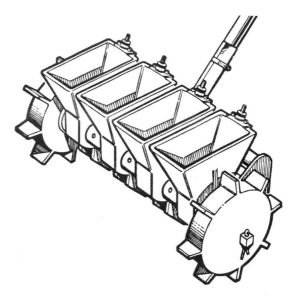

Fig 4.6   The French-made SM drill was used to sow four closely spaced rows of seeds at a time in cold frames and glasshouses.

Plate 4.6 Stanhay introduced their belt-feed precision seeders in 1953. The single-row hand drill was ideal for allotments and market gardeners could use two or more units on a garden tractor toolbar.

Plate 4.7   Hand-propelled fertiliser drill for lawns and the vegetable garden.

$1\frac{1}{2}$ in. apart or at any other distance equally divisible into 72. Pelleted seed was in common use by the late 1970s and this contributed to the improved accuracy of seed spacing in the row by the Exel and its many competitors.

Plate 4.8   A fertiliser spreader on a Bean self-propelled toolbar.

FERTILISER SPREADERS

A great deal of fertiliser was spread by hand in the late 1940s and this method was not limited to allotments and market gardens. Where space allowed, the alternative was a horse-drawn fertiliser distributor, probably made by Bamfords, International Harvester, Knapp, Massey–Harris, Nicholson or Tullos.

Many a ton of fertiliser was spread by hand from a bucket carried on a piece of rope slung around the neck during this period. Still more hand work was required on holdings where it was the practice to mix two or more straight fertilisers, with each containing a major plant nutrient. Compound fertilisers were available but many growers preferred to mix their own and save money.

Small hand-pushed fertiliser spreaders between 12 and 36 in. wide were made by Wolf Tools, Sisis and others in the 1960s for the application of lawn sand and fertiliser. Early spreaders had a grooved or notched wooden roller, with a small

Plate 4.9   The Vicon Varispreader on a Kubota tractor.

land wheel at each end, in the bottom of a small triangular shaped sheet steel hopper.

The 12 in. wide Fison spreader cost £2.68 in 1973 and a 24 in. Wolf spreader was £19.95. Application rate was varied by using different feed rollers and on some machines an adjustable shutter controlled the rate of flow from the hopper.

Plastics were used for hoppers and feed rollers in the late 1970s. The Fison 12 in. spreader with a polypropylene hopper and a shut-off lever cost £8.40 in 1978.

Broadcasters and full-width distributors, either towed or carried on the hydraulic linkage of compact tractors, were used to spread fertiliser on market gardens, golf courses and other large areas of grass.

Various models of spinning disc broadcaster and the Vicon Varispreader with an oscillating spout which distributed the fertiliser were used by smallholders and market gardeners in the

Plate 4.10   Agitator feed on the Warrick distributor was used to spread fertiliser and broadcast a wide variety of seed. It had a 7 ft 6 in. spreading width and the lightweight hopper could be tipped upside down for cleaning.

Plate 4.11   Hand-held syringes for controlling pests in the garden and small greenhouse.

1950s. Full-width fertiliser distributors, mostly with plate and flicker or reciprocating plate feed mechanisms, were also used but they were difficult to clean and many suffered from the corrosive action of fertiliser. Less complicated feed mechanisms, usually with rotating agitators spaced at intervals in the hopper and driven by the land wheels, appeared in the mid 1950s. Fertiliser was agitated through adjustable-sized outlets and deflector plates ensured even distribution across the width of the machine.

SPRAYERS AND DUSTERS

Hand-operated and horse-drawn sprayers have been used by gardeners and smallholders to apply pesticides, weed-killers and other chemicals to their orchards, fields and gardens for the best part of a century. By the mid 1940s nurserymen and market gardeners were finding

many uses for hand and garden tractor sprayers including lime washing, glasshouse shading, pest control and killing weeds. Domestic gardeners of the day were also gaining benefit from chemicals, usually mixed in a bucket and applied with a hand-operated syringe or stirrup pump.

Allman, Cooper Pegler, A & G Cooper, Dorman, Drake & Fletcher, Evers & Wall, Four Oaks and W. Weeks & Son were among a long list of companies who made sprayers in the 1940s and 1950s. Their products ranged from hand-held syringes for the home gardener to knapsack, barrow and garden tractor sprayers for market gardeners and fruit growers.

Hand-held syringes were sold by ironmongers and gardening shops in a variety of shapes and sizes during this period. Most were made from non-corrosive brass and the cheapest were sold for a few shillings. Syringes were filled by placing them in a container of chemical and

Fig 4.7   *A selection of Solo garden sprayers advertised in the mid 1950s.*

Plate 4.12　The contents in the Four Oaks Streetly hand sprayer were pressurised with the wooden handled pump.

withdrawing the plunger handle to suck liquid into the barrel. It was possible to spray almost non-stop when using more expensive hand syringes with a suction hose. This was placed in a container of diluted chemical and spraying could continue until it was empty.

The Four Oaks Streetly was a small hand-held pneumatic sprayer suitable for applying chemicals in glasshouses. An integral hand pump pressurised the 3 pint container, after which it was possible to spray non-stop until the tank was empty. The Streetly cost £5 12s 0d in 1959; a de luxe model with a thumb operated control lever was an extra 9s 0d.

E.J. Allman set up in business in a small shed as a motor engineer in 1919. He moved to a larger building at the company's present site a few years later and then added agricultural engineering to the business. The first Allman sprayer did not appear until 1946 but since then the Chichester company has become a market

Plate 4.13　The makers stated that a trial would soon convince potential buyers that the Allman Rapid sprayer was the most simple, efficient and easily manipulated sprayer available to the market gardener.

Plate 4.14   The Colwood motor hoe with an Allman sprayer.

leader with its range of knapsack, barrow and tractor spraying and dusting machinery.

The Pestmaster Minor dusting machine and the Speedesi powder duster patented in 1947 were the first Allman products. The Rapid sprayer, for weed control in row-crops, seed-beds, nurseries and sports turf, was first made by a Kent company in 1947 and the design was acquired by Allman in 1949. It had two brass pumps driven by a system of cranks from a bicycle wheel fitted with a freewheel, a six-gallon galvanised steel tank, a foot-operated on/off control lever and a pressure chamber cylinder to maintain constant working pressure. Three nozzles on both halves of the two section spraybar were spaced 12 in. apart and the bar could be set to a maximum height of 18 in. above the ground. The Rapid was also used with the two sections in a vertical position to spray fruit bushes. Application rates from 11 to 130 gallons per acre were possible by using different nozzles and varying walking speed. A powder duster version of the Mark I Rapid with the fan driven by a belt from the land wheel was also made.

The wheel driven Rapid was replaced with the Mark II sprayer with a J.A.P four-stroke engine and 6 ft spraybar in the early 1960s. The J.A.P engine was used for a year or so but by 1963 it had been replaced with an American-built $\frac{3}{4}$ hp Ohlsson & Rice two-stroke engine. The Mark III Rapid with a new design pump and a Villiers two-stroke engine which ran at 6,000 rpm cost £78 17s 6d when it was announced in 1966.

Allman introduced sprayer units for Landmaster rotary cultivators and the Colwood motor hoe in 1959. The Colwood unit had a $\frac{3}{4}$ in. roller vane pump with a maximum working pressure of 100 psi. It was vee-belt driven with a 5 to 1 speed reduction from a pulley on the engine crankshaft. The specification included an 8 gallon galvanised tank, pressure regulator, pressure gauge and a two-way on/off tap on the handlebars. The two-part spraybar could be used for vertical and horizontal applications and hand lances were provided for spraying trees, hedgerows, buildings, etc.

The Allman sprayer for Landmaster Gamecock and Kestrel garden cultivators used the same roller vane pump connected to the rotary cultivator drive shaft. A 20 gallon tank was close coupled to the tractor and carried on small castor wheels at the rear. Controls were similar to

Plate 4.15 The spray bar on the Allman Midget wet sprayer could be used either horizontally or vertically and two hand lances were provided for spraying fruit bushes and trees.

Plate 4.16 The Allman Plantector sprayer for smallholdings and market gardens was used on the Allis Chalmers Model B and similar small row-crop tractors.

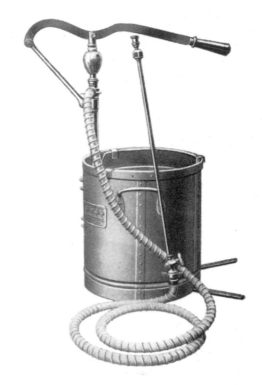

Plate 4.17   Lime sulphur, creosote and paraffin were among the liquids applied with this Four Oaks Knapsack sprayer, which cost £10 19s 6d in 1959. It had a 3½ gallon polished and lacquered copper tank, diaphragm pump and a brass hand lance.

Plate 4.18   The six-gallon Four Oaks Triumph bucket sprayer was supplied with 10 ft of hosepipe and a hand lance with different nozzles for limewashing and crop spraying.

those on the Colwood unit, hand lances were available and the spraybar could be used for vertical or horizontal spraying.

The Allman Gardenspray was originally manufactured in 1959 for Gardenmaster garden cultivators made by Landmaster but versions suitable for other small garden tractors with the rotor in front of the wheels were soon added.

The Gardenspray had a 2 gallon plastic tank, a brass piston pump, pre-set relief valve and a hand lance with an on/off control tap. The pump, which delivered up to 30 gallons per hour at a pressure of 40 psi, was driven by the cultivator rotor drive shaft. A hand lance with three different nozzles, which gave spray patterns from a fine mist to a pencil jet for winter washing fruit trees was supplied with the Gardenspray. Return flow from the relief valve was used to agitate the tank contents and keep them well mixed. The moving parts ran

in oil-impregnated bushes and no tools were required when fitting the sprayer attachment to the Gardenmaster.

Allman added the Midget Wet sprayer for the application of most types of insecticide, fungicide, herbicide and disinfectants to their range in 1963. A 1½ hp Villiers or 1¼ hp J.A.P four-stroke engine, which could be adapted to run on tvo at a small extra cost, was used to drive the pump. The first Midget sprayers had an Allman gear pump made from bronze but subsequent machines had a wear-resisting roller vane pump. The Midget barrow sprayer had a 6 gallon tank on a two-wheeled trolley and a 4 ft 6 in. spraybar with three low-volume nozzles. Spraying width could be increased to 8 ft 4 in. for sports fields and other large areas of land. Hydraulic agitation was used to keep the tank contents well mixed and operating pressure was in a range of 0 to 140 psi

Plate 4.19   The 18 gallon Four Oaks Stafford barrow sprayer cost £38 13s 6d. Pneumatic tyres were an extra £8 13s 6d.

depending on the size of the nozzles.

Allman may have looked into the future with their crystal ball, but more likely to meet the needs of continental growers, the Midget Wet sprayer could be supplied with a pressure gauge calibrated in psi or in bars.

The Four Oaks Spraying Machine Company at Sutton Coldfield in Warwickshire, specialists in spraying equipment since 1895, included the Excelsior, Kent, Stafford and Ely among the many different sprayers listed in their late 1940s catalogues.

The Excelsior pump for spraying, limewashing and disinfecting was used with a bucket or similar container. A hand-held plunger was used to draw the liquid into the pump body and deliver it through a length of hose to a single nozzle.

Pneumatic sprayers were already popular in the 1940s and 1950s. The Four Oaks Kent pneumatic knapsack sprayer, made in four sizes with 1, 2, 3 and 4 gallon tanks was a typical

example. A wooden-handled plunger pump was used to pressurise the brass container, which had a short length of hose connected to a brass hand lance. The four-gallon model weighed 18 lb and cost £17 11s 0d in 1959.

The Ely narrow-gauge barrow sprayer had an 18 gallon tank carried on a single-wheeled trolley with a hand operated pump, 15 ft of hose and hand lance. A sales leaflet suggested that the 12 in. wide Ely was ideal for spraying closely spaced row-crops.

The 1960 Four Oaks catalogue listed about 40 different knapsack, barrow and syringe sprayers. Many of them had galvanised steel tanks with brass pumps, hand lances and taps.

The Triumph was typical of the Four Oaks bucket sprayers with a 6 gallon container, brass pump, 10 ft of hosepipe and a hand lance. They were used to apply limewash, creosote, distemper, insecticides, disinfectant and other chemicals. A 1960 Four Oaks price list included the 4 gallon Farmer bucket sprayer at £8 5s 6d and the 10 gallon Rochester with an automatic agitator for £25.1s.0d.

The Four Oaks Colonial stirrup pump, which cost £5 7s 3d, had a $\frac{3}{4}$ in. diameter brass barrel, a 40 in. length of hose, nozzle and a polished wooden handle, was an even more basic spraying system. Chemicals, limewash, etc. were mixed in a suitable container and applied from the stirrup pump nozzle. The pump could also be used to extinguish a small fire.

Four Oaks also made hand-operated and engine-driven barrow sprayers in the early 1960s. The cheaper models were supplied with a hand lance while other more expensive sprayers had a hand lance and spraybar. The 6 gallon Victor barrow sprayer for row-crop work was similar to the single-wheeled Ely sprayer. It was the smallest model in the range with a double acting, hand lever operated brass pump and a 2 ft polished brass hand lance with a brass stopcock. The Bridgewater was the largest Four Oaks barrow sprayer at the time with a 25 gallon tank mounted on a two-wheeled trolley with handlebars to push it along. The Bridgewater, which cost £62 10s 0d on pneumatic tyres, was described as a most efficient and solidly constructed machine, designed for really hard wear under the most trying conditions.

The Four Oaks Senior sprayer, designed for market gardeners and large estates, had a Villiers 70 cc two-stroke engine to drive a gear pump, which supplied the three part spraybar or hand lances with chemical from a 15 gallon galvanised tank. The hand-propelled Senior cost £98 10s 0d but growers with some spare cash who were not keen to push the sprayer through their crops could buy a 25 gallon self-propelled Four Oaks Spraymobile for £148 10s 0d.

The pedestrian-controlled Spraymobile, introduced in 1960, had virtually everything except a seat. A Villiers four-stroke Mark 12 engine operated the pump and propelled the sprayer through a clutch and three speed gearbox. Two interconnected tanks with a total capacity of 25 gallons supplied a hand lance or a 10 ft spraybar with 7 nozzles.

Evers & Wall at Lambourn in Berkshire were another long-established farm and horticultural sprayer manufacturer making a full range of spraying equipment during the 1950s and 1960s. The Evers & Wall F.R. 2/RCS linkage mounted row-crop sprayer for the David Brown 2D,

Ferguson 20 and similar tractors, was according to sales information, a versatile machine which enabled one operator to do the work of six men. It was also claimed to commend itself to anyone who grew row-crops including market gardeners, nurserymen, soft fruit growers and seedsmen.

The F.R. 2/RCS had a 100 gallon tank and the pump could deliver 1,200 gallons an hour at pressures of up to 225 psi to various attachments suitable for all crops from seedling stage to 9 ft high trees. The horizontal spraybar was used with standard non-drip fan jets or swivel-mounted twin nozzles on tong-shaped drop legs (Plate 4.22). The swivel nozzles could be individually adjusted to give a spray pattern of the required shape, angle and volume. Fruit crops were sprayed with nine swirl jets on two vertical booms carried on an adjustable parallel linkage to suit row widths from 3 ft 6 in. to 9 ft. The

Plate 4.20 Although the Four Oaks Senior mechanised sprayer had an engine-driven pump, it was important for the operator to push the machine at a constant speed to ensure an even application of chemical from the spray bar.

Plate 4.21   The Four Oaks Spraymobile was a self-propelled version of the Senior mechanised sprayer.

Plate 4.22   Horizontal and vertical spraybars were supplied with the Evers & Wall F.R. 2/RCS row-crop sprayer.

Plate 4.23   The Multicrop sprayer made by Evers & Wall for the David Brown 2D had drop legs with the jets in adjustable swivel sockets.

spraybars remained upright even when set at their maximum height of 9 ft.

Knapsack sprayers, barrow sprayers and spray units for garden cultivators were also made by Evers & Wall in the 1960s. Knapsack sprayers included pneumatic and hand pumped models with 2 or 3 gallon brass tanks and hand lances working at 90 to 100 psi. The E & W 10 gallon barrow sprayer on rubber-tyred wheels had a powerful brass pump, 10 ft of hose and a hand lance with jets for lime washing and crop spraying.

Spraying attachments for the Howard-Clifford 700 and Howard Yeoman Rotavators, also Clifford Mark I and Mark IV garden tractors were another Evers & Wall product. They could apply up to 70 gallons an acre from the 7 ft 6 in. spraybar at a maximum pressure of 125 psi when travelling at 2 mph. The rate was increased to a maximum of 200 gallons an acre when spraying row-crops with twin jets on drop legs.

A sprayer kit was made by Evers & Wall for the Landrover and Austin Gipsy in the late 1950s. The tank, either one supplied with the kit or a suitable 40 gallon drum, was carried in the back of the vehicle and the spraybar was attached to the front bumper. The pump unit was fitted in place of the middle passenger seat, immediately over the vehicle's central power take-off shaft. The unit applied up to 70 gallons per acre at 4 mph and when it was not needed for spraying, the tank and pump could be used for washing vehicles and buildings.

A 50 gallon spraying attachment was also made in the late 1940s by the Dorman Simplex Sprayer Company for Land Rovers. Other Dorman products included barrow sprayers and special equipment for spraying potato crops with acid to kill the haulm before the crop was harvested.

The company changed its name to the Dorman Sprayer Company in 1951 and sprayer production continued at Ely until Ransomes acquired the Dorman sprayer range in 1980.

The pump on the 1950 Dorman 15 gallon wheelbarrow sprayer was driven by a 1 hp J.A.P four-stroke engine. The oil sump was modified to give an increased oil capacity, which according to a sales leaflet gave trouble-free running for prolonged periods. The spraybar covered

Plate 4.24 The sprayer unit for the Howard–Clifford 700 had the same type of drop leg and swivel socket as those used on the F.R. 2/RCS machine to give an infinitely adjustable spray angle.

Plate 4.25 Market gardeners who owned a Land Rover in the 1950s could use it to spray their crops with an Evers & Wall sprayer kit.

7 ft 6 in. and an anti-drip device on the pump was said to eliminate nozzle drip when the sprayer was shut off. It could be used for low or high volume application to turf and ground crops. The spraybar secured to the chassis by quick release clamps was simple to remove when changing to a hand lance.

A towed version with a stronger chassis, 15 ft spraybar and a drawbar for hitching it to a light tractor or Land Rover was added in 1952. When weed control was required in inaccessible places, the spraybar could be removed, connected to the pump with a suitable length of hose and carried by two people across the problem area.

Henry Cooper and Frederick Pegler founded Cooper Pegler, a general trading company in London WC2, in 1894. Simple insecticide sprayers were included in their stock and successful trading resulted in Mr. Pegler obtaining world marketing rights for the

French-made Vermorel sprayers.

First World War bombing forced a move to Surrey and later to Burgess Hill in West Sussex. Cooper Pegler was acquired by a Norwegian chemical and plastics manufacturer in 1972 and remained at Burgess Hill until 1992, when the company moved to Ashington in Northumberland.

Like other spray equipment manufacturers, Cooper Pegler made a wide selection of pneumatic and manually operated knapsack and barrow sprayers in the 1940s and 1950s. A breakthrough came in 1960 when Cooper Pegler were invited by ICI to manufacture the world's

first plastic knapsack sprayer. The resultant CP3 was the forerunner of a long line of Cooper Pegler knapsack models culminating in the introduction of the Series 2000 CP3 and CP15 sprayers with blow-moulded polypropylene tanks in 1994, one hundred years after the two partners started in London WC2.

The 3.2 hp Cooper Pegler Hurricane Minor and $4\frac{1}{2}$ hp Hurricane Major motorised knapsack sprayers, both with JLO engines, were introduced in the early 1980s. A series of optional accessories for the Hurricane made it possible to apply liquids at standard or ultra-low volumes, also chemical dusts and granules. The Hurricane

Plate 4.26 The Dorman Simplex barrow sprayer, which cost £76 10s 0d in 1951, could also be used with a hand lance or towed behind a tractor or vehicle. Two people could remove the spraybar from the chassis and carry it across sloping banks and other inaccessible areas.

Plate 4.27   Cooper Pegler pneumatic knapsack sprayer.

Plate 4.28   Cooper Pegler Eclair knapsack sprayer.

could also be converted into a flame gun for burning off weeds.

Spraying attachments were made in the 1950s for many of the more popular garden tractors and even the Allen Motor Scythe could be used to spray fruit trees and bushes. The Allen Noblox spraying attachment had a piston pump driven by the knife-drive crank, with suction and delivery hoses, pressure regulator and a hand lance. The pump was bolted to the cutter bar bracket and with the tank on a load-carrying platform above the engine or in a trailer, the Allen scythe became a self-propelled spraying unit.

Spraying equipment was included in the 1950 Clifford cultivator catalogue. The pump was attached to the rotary cultivator drive shaft and the 18 gallon tank was close coupled

Plate 4.29   An instantaneous spray control lance was supplied with the Cooper Pegler Ondiver pneumatic barrow sprayer.

under the tractor handlebars. Four jets were arranged around the tractor and a connection was provided for a hand lance. Control taps were fitted to the jets and with a single jet in use the pump pressure was between 350 and 400 psi, which was typical for horticultural sprayers at the time. When used with a single nozzle, the 18 gallon tank on the Clifford unit was emptied in 21 minutes. With four jets in use the pressure was reduced to 150 psi and the tank was empty after 6 minutes having travelled little more than 200 yards along a row.

Plate 4.30   A sprayer made for the Clifford Rotary cultivator was used to apply chemical from high- or low-level jets on both sides of the tractor or with a hand lance.

Plate 4.31   Spraying with an Allen Motor Scythe.

A similar outfit was made for the Allen & Simmonds Auto-Culto model M garden tractor in the early 1950s. The pump emptied the 40 gallon tank in 16 minutes and the 350 psi working pressure was sufficient to operate two hand lances. A team of three was able to spray fruit trees on the move with one person steering the tractor and the other two wielding hand lances.

The Coleby mobile sprayer was originally made as an attachment for Coleby garden tractors. It had a 30 gallon tank and a three-cylinder pump driven by a $3\frac{1}{2}$ hp J.A.P 4/3 four-stroke engine with an output of 120 gallons per hour. The maximum working pressure was 400 psi with a single nozzle and 250 psi when using a hand lance with two nozzles.

Small fruit orchards were usually sprayed with a knapsack sprayer or a hand lance and barrow sprayer or spray unit on a garden cultivator in the 1940s and 1950s. Permanent spraying installations with a large engine-driven pump and a system of underground mains with standpipes for hand lances were installed by many large-scale fruit growers in the 1930s and some were still in use 20 years later.

Tractor-drawn orchard sprayers were at an early stage of development in the 1950s and some of them required a brave soul to ride on the machine to direct the spray chemical on to the trees.

Clean Crops Ltd of Westminster announced a new version of their tractor-drawn Micron orchard sprayer at the 1953 Royal Show. A 420 cc B.S.A. engine was used to drive an impeller, which moved about 2,000 cu. ft of air per minute to atomise the chemical into droplets

Plate 4.32   The spray attachment for the Coleby Junior cultivator could be used with horizontal or vertical spraybars or hand lance.

Plate 4.33   The Coleby mobile sprayer, which cost £118 in the early 1950s, had a 60 gallon tank and a hand lance that could spray to the top of 25 ft high fruit trees.

of approximately 100 microns in diameter. A cloud of fine spray particles from a large nozzle on a flexible hose was directed on to the trees by an operator, who was required to sit on an uncomfortable-looking seat at the back of the sprayer. In ideal conditions the spray cloud travelled between 60 and 70 ft. The Micron orchard sprayer with a full 50 gallon tank and average sized operator weighed about 10 cwt and cost about £400.

Universal Crop Protection exhibited the Whirlwind trailed orchard sprayer at the 1952 Smithfield Show. It had an 88 gallon chemical tank on a two-wheeled trailer with a two-cylinder 10 hp J.A.P engine to drive the pump and centrifugal fan. The Whirlwind applied $1\frac{1}{2}$–2

Plate 4.34   Sprayer attachment for the Howard Bantam.

gallons of chemical per minute, which was atomised by a 180 mph air blast into 70–120 micron droplets. The spray mist was directed from a flexible hose and nozzle by an operator seated at the back of the machine. Sales information indicated that the atomised chemical had a vertical reach of 45 ft in still air and a horizontal reach of 60 ft with the help of a 5 mph wind. It did not mention how far the chemical would travel when the tractor was facing a head wind!

Most garden and horticultural sprayers were made of expensive but corrosion resistant copper and brass until 1960 when Cooper Pegler introduced a knapsack sprayer with a plastic tank.

Plate 4.35 The operator was required to direct chemical from the Whirlwind orchard sprayer pump to its target.

Apart from the hand lance, pump crank, harness buckles and a few screws the Cooper Pegler Policlair CP3 was made entirely from plastic materials by the mid 1960s. Polypropylene was used for the pump body and carrying harness, the four-gallon tank was made from polyethylene and the crankshaft bearings, pump piston and hose were nylon.

The Allman Polypak Kestrel knapsack sprayer introduced in the early 1960s was suitable for a wide range of chemicals, whitewash, creosote, etc. After initial pressurising with the diaphragm pump, a few strokes with the pump handle were sufficient to maintain the compression cylinder at its 65 psi working pressure.

Allman added the Minispray knapsack sprayer to their range in 1966. This was a lightweight model with a transparent $2\frac{1}{4}$ gallon plastic tank, piston pump and plastic hand lance which weighed a mere 6 lb when empty. Minispray attachments included a hand-held dribble bar applicator, the Arbogard for ring spraying around young trees and the Expando telescopic dribble bar with adjustable crop shields to cater for various row widths.

Sales of two-wheeled tractors, except for some domestic models, were in decline by the early 1960s but specialist companies were still making sprayer units for the Ransomes MG crawler, the Auto-Culto, Merry Tiller and a few other garden tractors. Auto-Culto International approved the

4.36 Operator comfort was considered in designing the Cooper Pegler Policlair knapsack sprayer. The tank was contoured to be a more comfortable fit on his or her back and according to sales information the hand pump was outstandingly easy to use.

1   The Kendall 8 hp tractor was made from 1945 to 1947.

2   The Mark 1 Trusty Steed was made from 1948 to 1950.

3   The American-built Allis Chalmers Model G was imported between the late 1940s and the mid 1950s.

4　The B.M.B President was made from 1951 to 1955.

5　Newman Industries bought the Kendall design and an improved three-wheeled Newman tractor was announced in 1948.

6　The first Uni-Horse tractors were made by Lea Francis in 1961.

7 The Maskell self-propelled toolbar was made at Wilstead in Bedfordshire during the late 1950s.

8 The four-wheel drive version of the Iseki TX1300 was imported by L. Toshi Ltd in 1976.

9 Massey Ferguson 1210, 1993.

10   Ransomes MG 2 cultivator and TS 42 plough, 1946.

11   The Bristol 25 replaced the 20 in the mid 1950s.

12   An early 1950s Trusty tractor.

13 Howard Gem, 1974.

14 Howard Dragon, 1979.

15 An Iseki KS 650 ridging up
potatoes in the early 1980s.

16   Kubota AT 55 Tiller, 1990.

17   Howard Rotavators in the mid 1960s showing the 200, 350, 400 and Gem Series V.

18   The Ransomes Vibro Hoe was made in the mid 1950s.

19 Two Stanhay seeder units on a BCS garden tractor in the late 1980s.

20 Spraying whitefly with a 1960s pattern hand syringe.

21 Allman barrow sprayer, 1989.

22 The Sprolley, 1994.

23 An early petrol-engined Flymo, the first hover mowers were made in 1964.

24 Flymo Turbo Compact 350 with an internal grassbox, 1994.

25   Hayterette rotary mower, 1966.

26   Hayter Harrier, 1994.

27   A mid 1950s Westwood rotary mower.

28   The rear grassbox version of the Qualcast Concorde, 1980.

29   The 1980 E 30 de luxe version of the electric Qualcast
Concorde with a front grassbox. The first Concorde mower
was made in 1971.

30   A late 1970s model of the Qualcast Jetstream
introduced in 1971.

31   The mains electric Qualcast Mow &
Trim was launched in 1984.

32   Production of the Ransomes 24 cylinder mower continued after the Ipswich company stopped making domestic mowers in the mid 1970s. The last Ransomes 24 mowers were made in 1981.

33   Bertolini cutter bar mower, 1994.

34   Ariens Bagger Vac mower on an Iseki TX 2160, 1986.

35   This John Deere mower is typical of ride-on machines made in the
late 1980s.

36   Mid-mounted mower on a Kubota G3HST compact tractor, 1987.

37   The Opperman Motocart was popular in the mid 1950s.

38   Kubota B1750 fitted with turf tyres hitched to a trailer, 1990.

39   Kubota tracked carrier, 1994.

40   Wrigley Union motor truck, 1994.

41 A 1994 Allen Leaf sweeper, the first machines were made in 1972.

42 Black & Decker lawn raker, 1985.

use of the Allman Autospray with the Auto-Culto 65, Mark IX Autogardener and Midgi-Culto and Wolseley recommended the Allman Merryspray for the Merry Tiller.

The Autospray and Merryspray introduced in 1959 had roller vane pumps vee-belt driven from a pulley on the engine crankshaft with an output of $2\frac{1}{2}$ gallons per minute at a maximum pressure of 150 psi. Chemical from the $5\frac{1}{2}$ gallon plastic tank on the handlebars was applied by three jets on each half of the two-section spraybar at rates of 11 to 130 gallons per acre. The spraybar covered 6 ft when working in row crops and the two halves could be used in a vertical position for spraying fruit bushes. A hand lance for orchard work and a pressure washing gun increased the versatility of the Autospray and Merryspray.

A & G Cooper, Drake & Fletcher, Ransomes and Weeks made sprayers with power take-off driven pumps for the Ransomes MG market garden tractor. Ransomes own Mark I Crop-guard Junior had a 30 gallon tank carried on the MG's hydraulic linkage with a nylon roller vane

Plate 4.37 The Allman multi-purpose Kestrel Polypak knapsack sprayer, with a choice of two- or three-gallon translucent polythene tanks, was launched in 1961.

Plate 4.38 The Allman Autospray was approved for use with Auto-Culto garden tractors.

Plate 4.39   Allman produced the Merryspray for the Wolseley Merry Tiller.

Plate 4.40   The pump on the Weeks Model M sprayer for the Ransomes MG crawler had an output of 250 gallons per hour at 200 psi pressure.

pump, a 17 ft spraybar and optional hand lances for orchard spraying.

Trailed sprayers were made for the MG by A & G Cooper at Wisbech, W. Weeks & Son at the Perseverance Works in Maidstone and Drake & Fletcher, also of Maidstone in Kent. The 100 gallon Cooper Demon Z200 pump delivered $3\frac{1}{2}$ gallons per minute to two hand lances at a maximum pressure of 400 psi. The pump was mounted on brackets bolted to the tractor and driven by a roller chain from the power take-off.

The Drake & Fletcher L.O. Estate sprayer had a 60 gallon tank and the pump was mounted on a platform attached to the tractor and driven from the power shaft by a belt. The pump delivered up to 6 gallons per minute at 250 psi and a hydraulic agitator kept the tank contents well mixed. The L.O Estate sprayer could be used with a 12 ft spraybar, hand lances, adjustable nozzle guns for orchard spraying or a special spraybar with 15 jets for hop gardens.

The Weeks Model M sprayer pump, driven by a universal drive shaft from the power take-off, delivered 4 gallons per minute at 200 psi. The pump and 50 gallon tank were mounted on a tubular two-wheeled chassis running on pneumatic tyres and the Model M could be used to spray ground crops or soft fruit and hand lances were provided for orchard work.

Knapsack and wheelbarrow sprayers were included in the Dorman Sprayer Company's 1961 catalogue. The Dorman wheelbarrow sprayer had a 15 gallon steel tank and a bronze and stainless steel gear pump driven by a J.A.P four-stroke engine. It cost £75 complete with 7 ft 6 in. spraybar and five anti-drip nozzles. An optional 10 ft 6 in. spraybar was an extra £5 10s 0d and a hand lance with 15 ft of hose cost £4 17s 0d.

By the mid 1960s the Dorman range included

Plate 4.41 Orchard spraying with the Cooper Demon sprayer was done with 6ft hand lances but 3 ft lances were advised for bush fruit and glass houses.

Plate 4.42 This Dorman wheelbarrow sprayer with a 15ft spraybar and 25 gallon tank cost £115. It could be pushed or fitted with a towbar and pulled by a Land Rover or similar vehicle.

the Osprey Chick plastic tank knapsack sprayer for domestic gardens priced at £3 3s 6d and the larger 2 gallon Osprey Continental was £6. There were three models of Dorman wheelbarrow sprayer comprising the 10 gallon Ely and Wheelaway with engine driven pumps and the Junior Pneumatic sprayer with a 2 gallon tank.

The two-wheeled Ely with a 7 ft 6 in. spraybar was mainly used on sports fields and the single bicycle wheeled Wheelaway with six jets spaced across a two-part spraybar was designed for market garden work. An improved Wheelaway II barrow sprayer had a 5 gallon plastic tank on both sides of an auto-cycle type wheel with a 75 cc four-stroke engine to drive a

gear pump. This delivered chemical to a 9 ft, two part plastic spraybar at a rate of $3\frac{1}{2}$ gallons per minute.

Plastic became the standard material for many sprayer components in the early 1960s and in the same way, diaphragm and vane pumps had replaced the wear-prone gear pump by the late 1960s. Allman, Dorman, Evers & Wall and ICI Plant Protection were among the sprayer manufacturers who used this new technology.

The ICI Plant Protection used plastics for the Polyrow barrow sprayer. Designed for inter-row work it had a $2\frac{1}{2}$ gallon plastic tank, ground wheel driven pump and a single flood jet with two spray guards to protect the rows of plants.

Plate 4.43   The Dorman Junior pneumatic sprayer tank was pressurised with a foot pump supplied with the machine. One tankful of chemical was sufficient to treat $\frac{1}{4}$ acre of fine turf. The Junior cost £32 10s 0d ex works in 1966.

The Hardi three-wheeled barrow sprayer made by Evers & Wall had a 44 gallon plastic tank and twin diaphragm pump driven by a Briggs & Stratton engine. Dorman barrow sprayers in the mid 1970s included the Auto-Spraya and the Osprey Wheelaway, a slightly modified version of the earlier Wheelaway II sprayer. The single-wheeled Auto-Spraya with a 4 gallon plastic tank and 6 ft spraybar was designed for spraying fine turf. A lever type pump operated by the wheel delivered chemical at rates of 10 and 15 gallons per acre. Dorman's 1978 catalogue included a barrow model for orchard work with a 22 gallon tank and hand lances priced at £520 and the 10 gallon Ely Groundsman with a Briggs & Stratton engine cost £450.

The Allman Rapid barrow sprayer first seen

Plate 4.44 A pressurised knapsack container supplied chemical to the jets on this 1960s Dorman band sprayer.

Plate 4.45 The Allman Rapid Mark V barrow sprayer, which cost £397 in the mid 1970s, had a four-stroke engine-driven twin diaphragm pump to supply a spraybar with six nozzles.

Plate 4.46   Allman Arbogard sprayer for ring weeding young trees.

Plate 4.48   A hand lance was standard equipment on the 15 gallon Allman CR 70 barrow sprayer. A 9 ft spraybar was optional.

Plate 4.47   This ASL compression sprayer cost £4.25 in 1973 but by 1978 inflation had more than doubled the price to £11.53.

some thirty years earlier had reached the Mark 5 version by the mid 1970s. The specification included a 12 gallon plastic tank, 12 ft spraybar, hand lance and a twin diaphragm pump driven by a Briggs & Stratton 3 hp engine or an electric motor.

Allman introduced a new four-wheeled trolley-mounted orchard sprayer with a 55 gallon plastic tank, two hand lances and the Midget portable sprayer unit in 1972. The Midget, introduced in the mid 1960s, was a self-contained unit with a four-stroke engine, pump, relief valve, pressure regulator and gauge with two sets of hoses and hand lances. The first Midget spray units had a 75 cc Villiers petrol engine but later models were supplied with a 3 hp Briggs & Stratton engine or an electric motor.

The Arbogard Mark B1, launched by Allman

in 1969, was used with a knapsack sprayer to ring-weed young trees with Gramoxone. The Arbogard—an ICI trademark—had a single flood jet and an adjustable plant guard to protect small trees. A larger Mark II Arboguard with two jets for commercial orchard and forestry work was added in 1972.

Domestic gardeners were offered a wide choice of pneumatic and small hand-held sprayers during the 1970s. A compression sprayer with a 1 gallon plastic tank cost between £5 and £10 and there was change from a one pound note for a $\frac{1}{2}$ pt trigger-operated mist sprayer.

Many manufacturers noted for their spraying equipment in the 1960s and 1970s had been taken over, ceased trading or given up horticultural sprayer production by the mid 1980s. However, a few of them, including Allman and Cooper Pegler, were still making crop protection equipment for gardeners and commercial horticulturalists.

The 1983 Allman range included the Rapid Mark V, a 55 gallon trolley-mounted unit and knapsack sprayers with plastic tanks. The Rapid sprayer, first made in 1947, finally went out of production forty years later in 1987 when Allmans announced the new CR70 and CR100 barrow sprayers. The 15 gallon single-wheeled CR70 and the 22 gallon CR100 two-wheeled barrow sprayers have diaphragm pumps driven by a 3 hp Briggs & Stratton engine. A hand lance with 25 m of hose on a convenient reel is standard equipment on both models and a 9 ft spraybar is an optional extra.

Plate 4.49 A late 1980s Walkover Groundsman sprayer for fine turf.

Cooper Pegler's products in the late 1980s included the Falcon pneumatic sprayer with galvanised or stainless steel containers and CP Knapsack sprayers with a plastic tank and lever-operated diaphragm pump. Two-wheeled barrow sprayers had given way to three-wheeled trolley sprayers with a 1½ hp two-stroke petrol engine or an electric motor to drive the diaphragm pump. Other features of Cooper Pegler trolley sprayers include a 20 gallon fibreglass tank, a hand lance and length of hose and a two-part 8 ft wide spraybar. The sprayer could be towed with an optional drawbar after removing the single front wheel.

Mounted sprayers made in the 1950s for the Ferguson 20, Allis Chalmers B and similar tractors were used by market gardeners and smallholders. Most farm sprayers were too large for this type of work by the late 1960s and this led to the introduction of a new generation of small mounted sprayers for compact tractors. Soon after Lely Import became Iseki distributors in 1982, they introduced the Conquest 88 sprayer for Iseki, Kubota and other compact tractors. The Conquest with an 88 gallon plastic tank had many of the features found on farm sprayers at the time, including a twin-diaphragm pump, stainless steel booms and quick fit, colour-coded jets.

The Walkover sprayer, developed by Paul Ridgeon for the application of weed killers and liquid fertilisers to lawns, was launched in 1980. A land-wheel driven pump delivered chemical from a small plastic tank to jets placed near the ground at the front of the machine. The pump only supplies chemical to the jets when the sprayer is pushed in a forward direction and flow ceases when the machine is stationary or pulled backwards. Allen Power Equipment acquired Walkover sprayers in 1990 and since then the range has grown from a simple sprayer for domestic lawns to include larger models for gardens, plant nurseries and sports grounds. The Fieldmaster trailed sprayer for ride-on tractors with a 12 gallon tank has an output of about 14 acres per day with the standard 6 ft spraybar.

A dual-purpose boom and hand lance sprayer, which may one day replace the knapsack sprayer, was introduced in 1994. Made by L.E. Toshi Ltd, the original importer of Iseki compact tractors, and now marketed by Sisis, the Sprolley is a sprayer on a lightweight hand trolley. It has a diaphragm pump driven by an electric motor supplied with current from a 12 volt battery, which also provides electricity for the powered front axle on the self-propelled model.

CROP DUSTERS AND MIST BLOWERS

Insecticide and fungicide powders and dusts have been used to treat horticultural crops at very low rates for many years. Various types of powder applicator or duster, which issued a penetrating cloud of fine dust, were made in the 1940s and 1950s. They ranged from small hand-held dust guns for gardens and greenhouses to dusting attachments for garden tractors and most of them were apparently made with scant consideration for the user's health and welfare.

A 1948 Ransomes sales leaflet explained the benefits to be gained from applying insecticide dusts with the Ransomes flea beetle duster to control flea beetle and turnip fly on brassica seedlings. The duster was pushed by hand and a land-wheel driven roller feed mechanism applied the dust in a 6 in. wide band over each row of young plants at a rate of 35 to 45 lb per acre.

Hand-operated dusters for gardeners and smallholders were made by several companies including Allman, Drake & Fletcher and Cooper Pegler. The Drake & Fletcher New Armada Mistifier, which cost £7 19s 0d in the mid 1950s, was carried on the operator's chest and the dust was blown through a single or double outlet by a stream of air from a hand-cranked fan.

The Cooper Pegler Carpi duster, with hand-

Fig 4.8 *Ransomes flea beetle duster.*

Fig 4.9   Drake & Fletcher Armada dust gun.

operated bellows, cost £2 17s 6d in the late 1960s. The user was required to pump the hand bellows, which blew the dust through a spreader nozzle at the end of a 3 ft tube while walking along a row of plants.

Plate 4.50   The Allman HandOp dusting machine with a hand-cranked six-bladed fan, introduced in 1946, was strapped to the operator's chest.

The Allman HandOp hand-cranked dusting machine, introduced in 1946, was carried on the user's chest with fully adjustable shoulder straps to hold it in place. The HandOp weighed 21 lb with a full hopper of powder, which was distributed from a nozzle by an air blast from a hand-cranked, six-bladed fan. Sales literature stated that the HandOp would dust ground crops, bushes or trees with equal efficiency and that it was light, strong, perfectly balanced and easy to operate for long periods without undue fatigue. The technical details explained that the 55 rpm fan had an output of $47\frac{1}{2}$ cu. ft of air per minute at a velocity of 2,500 ft per minute!

The Allman Speedesi, the first powder duster made by Allman, was awarded a Silver Medal for new implements at the Royal Agricultural Society's Show held at Lincoln in 1947. The Speedesi could be fitted to various farm and garden tractors including the Ransomes MG2 and the engine exhaust gases were used to distribute powdered pesticides on to the crop. As well as carrying the powder to the spreader outlets, the exhaust gas was also used to preheat the dust before ejection, which sales literature pointed out took full advantage of the well-known fact that warmed powder was more effective in the extermination of pests. The Speedesi could be used to apply dust to ground crops including potatoes and celery from outlets spaced across a 10 ft wide boom from pendant nozzles and in orchards with two fixed fish-tail nozzles or a long-handled dusting lance.

The Allman Pestmaster Minor duster had a single bicycle-type wheel, tubular handlebars, powder hopper, feed mechanism and a

Fig 4.10   Sales literature described the Allman Pestmaster Minor duster, made between 1946 and 1960, as an all-in-one machine suitable for insecticides, pesticides and powder fertilisers with application rates adjustable to meet all requirements.

Plate 4.51 This late 1940 Allman Speedesi duster, working in fruit with two fish-tail outlets, would in no way conform with 1990s application of pesticides regulations.

Plate 4.52 Dusting fruit trees with the Speedesi hand lance.

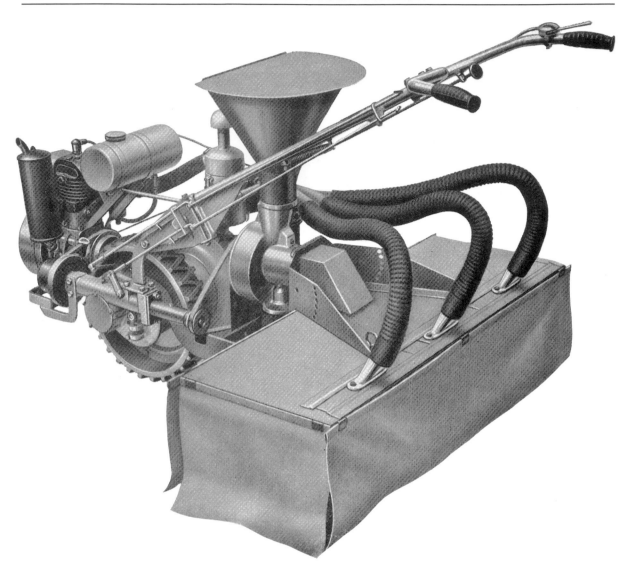

Plate 4.53   The nozzles create an intense cloud of dust under the canopy on the Howard Bantam recirculation duster.

centrifugal fan with delivery pipes. Two small brushes in the hopper agitated the powder and swept it at a controlled rate through an adjustable-sized outlet into an air flow created by a wheel-driven fan. The powder was blown from the delivery pipes through fish-tail spreaders or curved deflector plates.

The Allman Midget duster, with a 34 cc two-stroke engine and fan, was attached to straps on the operator's chest and the powder hopper was carried on his or her back. The fuel tank contained sufficient petrol and oil mixture for $1\frac{1}{2}$

hours work and the hopper held 20 lb of dust. An alternative model required the operator to strap the engine, fan and 12 lb capacity powder hopper on to his chest. The more enthusiastic person could dust two rows at a time by attaching a two-way branch and twin flexible dusting tubes to the fan housing. Allman also made a barrow version of the Midget duster with one or two wheels and handlebars.

The Drake & Fletcher Motorised Dust Gun had a crankshaft-mounted fan sandwiched between the 6 lb capacity powder hopper and

Plate 4.54   A brush-feed mechanism metered the powder from the 24 lb capacity hopper into the air stream of this Allman duster on the Farmers' Boy light tractor.

two-stroke engine. The dust gun with a full hopper weighed about 24 lb and the operator who was required to carry the machine and direct the nozzle at the target, could apply up to 60 lb of powder per acre.

Dusting attachments were made for a number of different garden tractors in the 1950s. The Recirculation Duster for the Howard Bantam had a centrifugal fan driven by the cultivator rotor power shaft. The powder was released into an air stream from the fan and discharged from three nozzles under a canvas canopy to form an intense cloud of dust. Some of the dust adhered to the plants and the rest was sucked back into the fan unit housing, where it mixed with additional powder supplied from the hopper. The fan was vee-belt driven and the feed agitator in the hopper was driven by a worm and gear wheel arrangement. Application rates of 40 to 120 lb per acre were possible and the amount was regulated by selecting any one of four worm wheels to drive the powder agitator.

Potentially health-threatening hand-operated dusters were replaced with combined dusting and misting equipment in the mid 1950s. The new equipment was engine driven and carried knapsack fashion on the operator's back.

The Drake & Fletcher Mamba could be converted from a mist sprayer to a duster in a minute or so by swopping the tank for a dust hopper. A 75 cc engine with a crankshaft-mounted fan was used to apply the mist at rates of a few pints per acre to control pest infestations or up to 20 gallons per acre in orchards. The spray mist or powder was discharged from a hand-held flexible hose and nozzle and directed on to the target by the user.

Kent Engineering & Foundry made the KEF Motoblo combined knapsack mist blower and duster in the mid 1970s. It weighed 53 lb with its $3\frac{1}{2}$ hp engine and full $2\frac{1}{2}$ gallon polythene tank and could apply up to six pints per minute from the hand-held discharge nozzle. The smaller Motoblo Junior with a 2 hp engine also applied up to 6 pints per minute and with a full tank it weighed 36 lb.

4.55 Converting the 1983 model Allman L 80 knapsack mistblower/duster from dusting to mist blowing was a simple task. The 77 cc two-stroke engine-driven fan had a range of 55 ft in still air.

The 1970 Allman catalogue included a knapsack Mistblower/Duster, which with a full range of accessories could be used for misting, dusting and flame weeding. The basic price was £65 but by 1977 inflation had increased the price of the Allman L80 knapsack Mistblower/Duster to £185 and the flame thrower for killing weeds was still included in the list of optional extras.

Equipment for applying pesticides at rates as low as 1 pint per acre became available in the mid 1950s. The principle of ultra low volume spraying with a spinning disc to atomise liquids into minute droplets was developed by Edward Bals, who founded Micron Sprayers in 1953. It was found that tiny quantities of liquid, evenly atomised and distributed, controlled pests equally as well as more traditional application rates with conventional sprayers.

The development of new lightweight plastics took the principle of very low volume application a stage further in 1970 with the introduction of a hand-held Ultra Low Volume Applicator. The Ulva sprayer applied 70 micron droplets from a small spinning disc driven at 6,000 rpm by a $1\frac{1}{2}$ volt electric motor. When spraying outside, the wind was utilised to carry the minute droplets of chemical to the plants but the more

Plate 4.56 The hand-held Micron rotary atomiser, driven by a $1\frac{1}{2}$ volt electric motor, applies pesticide at a rate of 1 pint per acre.

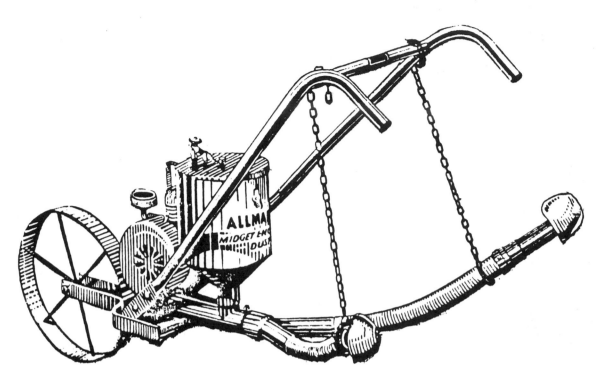

Fig 4.11 *The Midget duster, shown in the 1948 Allman catalogue, could be used in knapsack form or mounted on a single-wheeled chassis.*

expensive Ulvafan was required for glasshouse crops. This hand-held unit had an electric motor driven fan behind the atomiser unit to create the necessary air flow to carry the chemical to its target.

Other types of ultra low volume applicator produced during the mid 1980s included a hand-held twin atomiser with a knapsack chemical container and the Micronette barrow applicator carried on a single wheel with a 1 gallon tank and a small petrol engine to drive the fan and atomiser disc.

5 Lawn Mowers

Except for the gardens around stately homes, mansions and in public parks, a closely trimmed lawn was a rare sight until the late 1800s. Lawns were usually cut by teams of three or four men with scythes, a task requiring considerable skill to achieve a good finish. Daisy heads were chopped off between mowings with a daisy rake. Although Edwin Budding made the first lawn mower in 1830 it was many years before they became commonplace throughout the land.

Although this book is mainly concerned with the history of lawn mowers since the mid 1940s, the story would not be complete without a brief account of the development of grass cutting machinery since the 1830s.

Budding was an engineer responsible for the machines used to shear the nap from cloth in a textile factory and he conceived the idea of using a similar principle for cutting grass. He went into partnership with John Ferrabee at Stroud in Gloucestershire and the first Budding-designed mowers were made at his Phoenix foundry in 1831. Two years later, licences were granted to J.R. & A. Ransome at Ipswich, Thomas Green at Leeds and Alexander Shanks in Arbroath to manufacture the Budding Mower. It was gear driven from the rear roller with a flat box at the front to collect the grass cut by the 21 in. wide cylinder. Ransomes improved the design from time to time and by 1852 about 1,500 machines had been made at Ipswich.

Edwin Budding suggested that country gentlemen might find using his machine themselves to be an amusing, useful and healthy exercise. Whether the gentlemen actually found this to be the case when pushing a 21 in. cut mower must be open to question.

The first side-wheel mowers were made by Follows & Bate in the late 1860s. They were cheap to buy and the four models with 6 to 10 in. wide cutting cylinders soon became popular with the gardening public.

Fig 5.1 *This advertisement for a Qualcast side-wheel mower appeared in 1938.*

Plate 5.1 J.R. & A. Ransome were granted a licence to manufacture Edward Budding's gear-drive lawn mower at Ipswich in 1833.

Pennsylvania side-wheel mowers, made at Philadelphia in America, were sold in Britain by Lloyd Lawrence & Co. of London EC2 who later became Lloyds of Letchworth in Hertfordshire. The Vicar of Booton in Norfolk, one of the many Pennsylvania hand mower owners in the early 1900s, wrote to the Company complimenting them on their machine, which had exceeded all expectations. He added that his man could use it with only one hand when he pleased.

In 1841, Alexander Shanks made a 27 in. lawn mower light enough for a small pony to pull without damaging the grass with its hooves. The phrase 'to travel by Shanks' Pony' is said to be derived from the days when people walked behind a Shanks pony mower. A 42 in. cut horse-drawn mower was developed from the original machine a year or two later and mowers for animal draft were made for about ninety years.

A horse-drawn 'Horse Power' lawn mower with cutting widths from 30 to 48 in. for large

Plate 5.2   Features of the Follows & Bate S.2 12 in. cut side-wheel mower included a handle and roller of finest quality varnished wood, 7 in. diameter wheels with wide rims to ensure a good grip and a cutting cylinder with five Sheffield steel spiral blades.

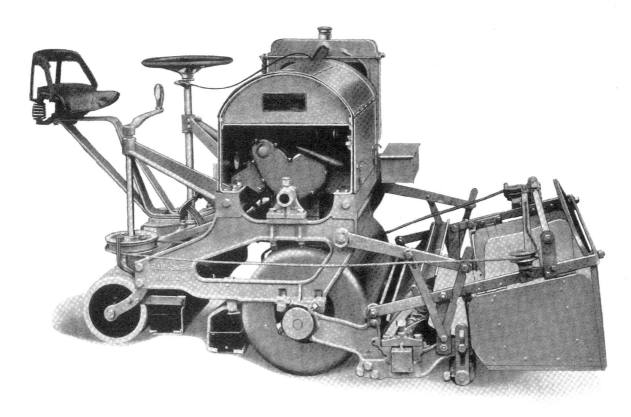

Plate 5.3   This Ransomes, Sims & Jefferies 42 in. cut ride-on mower with an 11½ hp water-cooled engine cost £375 in 1934.

lawns, cricket grounds and tennis courts was shown in Ransomes, Sims & Head's 1870 catalogue. Pony mowers with 26 and 30 in. cutting cylinders were added a year or so later. The horse wore leather boots to minimise damage to the turf. A set of four boots cost £1 5s 0d in the late 1890s.

The first steam-powered lawn mower, weighing 1½ tons, was made by the Lancashire Steam Motor Co. Ltd in 1893. After getting up steam, apparently in ten minutes from cold, the pedestrian-controlled mower with a grass collector was ready to go. The Lancashire Steam Motor Co. eventually became Leyland Motors Ltd. Seven years later Alexander Shanks made a steam-driven ride-on roller mower claimed to be more economical than animal draft and offered a complete absence of marks from horses' hooves.

Ransomes, Sims & Jefferies sold the world's first commercially manufactured petrol-engined motor mower in 1902. Designed by James E. Ransomes, the 42 in. cut cylinder mower with a 6 hp water-cooled engine enjoyed considerable success. Within a year Ransomes were also making 36 and 42 in. ride-on mowers and 2½ hp pedestrian-controlled models with 24 and 30 in. cutting cylinders. Thomas Green of Leeds was not far behind James Ransome with his 24 in. motor mower in 1903. Ransomes had sold more than 600 motor mowers by 1914 in spite of stiff competition from several other companies.

Charles H. Pugh made the first domestic motor mower in 1921 and the next twenty years saw the rapid development of small pedestrian-controlled mowers. Many companies were involved in the quest to build better and more efficient models including Atco, Dennis Brothers, Greens, J.P. Engineering, Presto, Royal Enfield, Qualcast, Ransomes and Shanks.

Electricity was an alternative power source

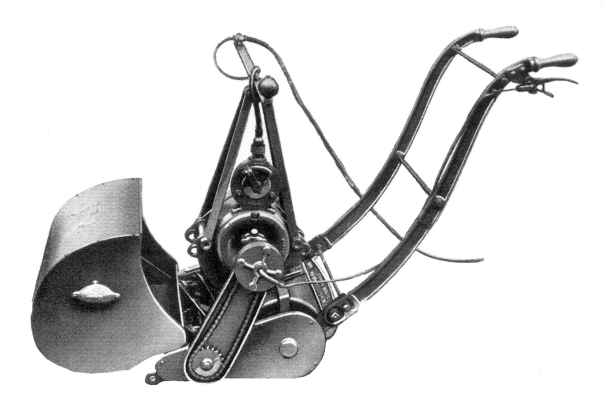

Plate 5.4   The 14 in. cut Ransomes Electra with 1½ hp mains electric motor cost £27 10s 0d in the mid 1930s.

Plate 5.5   A 1.9 hp Villiers engine was powerful enough to propel the Allen scythe with an operator on the trailer seat and cut a 3 ft wide swath.

and Ransomes claimed another first in 1926 with the launch of the Electra, a mains electric cylinder mower for domestic use, followed by the Bowlic for bowling greens a year or so later. Webbs introduced a mains electric mower in 1947, their first battery electric model appeared in 1959.

Side-wheel, roller-drive and motor mowers were made in the 1930s by a number of manufacturers including Arundel, Coulthard & Co. (Presto), Dennis Brothers, J.P. Engineering, Charles H. Pugh (Atco), Qualcast, Suffolk Iron Foundry and H.C. Webb.

Some of these companies were still in business by the late 1940s when lawn mower development was mainly concerned with the use of better materials and improved engineering design. Many of the thousands of side-wheel and roller mowers made in the late 1930s were still in use twenty years later.

The scissor cutting action of reciprocating knife mowers had been used in agriculture for almost a century before it appeared on the horticultural scene in the 1930s. The self-propelled Allen scythe, made for nearly forty years, was by far the best known. Atco, Lloyd, Mayfield, Teagle and others made hand-propelled machines with an engine-driven cutter bar and attachments were available for some two-wheeled garden tractors.

Power Specialities Ltd of Maidenhead made the world's first rotary mower in the 1930s, the company was bought by J.E. Shay Ltd in 1952 who in turn became part of Wolseley Engineering in the early 1960s. Hayters of Spellbrook, a name synonymous with rotary mowers, entered the domestic lawn mower market in 1947 with the launch of their 24 in. Motor Scythe and Vivian Loyd of Camberley introduced the Motor Sickle in the late 1940s. Both machines were rather dangerous as there were no guards over the vee-belt drive and the high speed cutting blades, which resulted in stones and debris being thrown in all directions.

The revolution in domestic rotary lawn mower design in the early 1950s coincided with the development of efficient, lightweight vertical crankshaft engines. Another milestone was reached in 1964 with the introduction of the heavily patented petrol-engined Flymo rotary hover mower.

Hundreds of different mowers have been made since 1945 and the following pages can provide little more than a snapshot of the manufacturers and the machines they have made during the last fifty years.

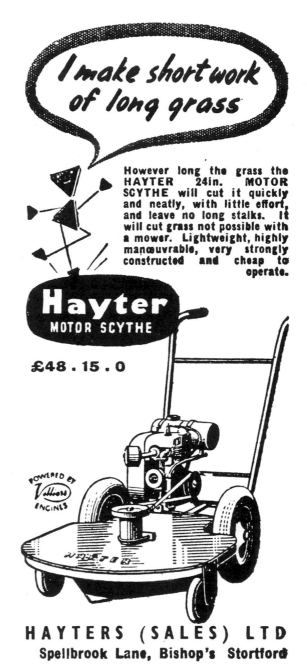

Fig 5.2 An early advertisement for the Hayter Motor Scythe.

CYLINDER MOWERS

Hand-pushed, roller-driven and side-wheel cylinder mowers were made in their millions in the late 1930s. Manufacturers included Atco,

Dennis, Follows & Bate, Greens, Presto, Qualcast, Ransomes and Webb. Many of those pre-1939 machines remained in use until the late 1940s when lawn mower production returned to normal.

The history of Qualcast can be traced back to 1801, when stoves and grates were made in Sheffield by the Jobson Foundry Company. The foundry business moved to Derby in 1849 and shortly after it was renamed the Derwent Foundry Company. The first Qualcast lawn mowers were made in 1920 at Derby and before long the company was selling a range of side-wheel and roller mowers at prices below most other British-made machines. The Derwent Foundry was a major lawn mower manufacturer by 1928, when it became a public company and changed its name to Qualcast, a derivation of Quality Castings.

Qualcast acquired Follows & Bate in 1938 but the onset of war brought lawn mower production to a temporary halt. Munitions including mortar bombs and hand grenades were made at Derby during the war years.

The Suffolk Iron Foundry was established by Mr L.J. Tibbenham at Stowmarket in 1920 to manufacture castings for agricultural and electrical equipment. Five years later Suffolk Iron Foundry added a range of domestic items including lawn mowers, line markers and tennis posts, Fleetway wringers, kitchen furniture and Clipper hand drills. Further expansion came in 1930 with the introduction of Sifbronze oxy-acetylene welding equipment. Tank transporters and bomb trolleys were made at Stowmarket during the war years and when the factory returned to peacetime production hydraulic pumps for farm tractors were added to the product range.

The Suffolk Colt, the first motor mower to be made at Stowmarket, was introduced in 1954. Four years later in 1958, Suffolk Iron Foundry joined the Qualcast Group and the acquisition of Charles H. Pugh Ltd in 1962 added Atco mowers to an expanding product range.

Charles H. Pugh started a wholesale jewellery and ironmongery business at Rotherham in 1865. Jewellery was dropped in favour of engineering and the company eventually moved to Birmingham. Interest in lawn mowers was kindled in 1919 and a notable first was achieved in 1921 with the introduction of the first petrol-engined, domestic motor mower.

The Qualcast Group with the Atco, Qualcast and Suffolk brand names merged with Birmid Industries in 1967 to form Birmid Qualcast. Atco and Suffolk Iron Foundry amalgamated in 1969 to become Suffolk Lawn Mowers and Atco mower production was moved to Stowmarket in 1975.

Birmid Qualcast widened its interests in 1984 with the acquisition of Wolseley Webb and Landmaster. Birmid Qualcast was bought by Blue Circle Industries in 1988 and more recent history has seen the closure of the Derby factory in 1991 and mower production concentrated at Stowmarket. Blue Circle disposed of its garden products division in a management buyout in 1992 and the new owners adopted Atco Qualcast as the company name.

The first hand-pushed Qualcast Panther roller mowers were made at the Victory Road factory in Derby in 1932 and within six years more than a million of them were in use. Following the war period, limited production of the 12 in. cut Panther was re-started at Derby in 1946. A 1950 magazine advertisement informed readers that the self-sharpening Panther, which ran on ball bearings and had a simple 'click' adjustment for the cutting cylinder, was the world's most popular mower with more than 3 million satisfied users.

A new 12 in. cut Qualcast Model B1 side-wheel mower was introduced in 1949 with a light tubular steel handlebar, which was a somewhat revolutionary change from the traditional Tee-shaped varnished wooden handle. Solid rubber cross-tread tyres were another innovation claimed to give a perfect grip with a silent and easy action, which reduced friction to a minimum. The 14 in. cut H1 side-wheel mower, which cost £5 19s 6d, was added to the Qualcast range in 1951. It had solid rubber-tyred wheels and a tubular steel handle with rubber handgrips.

Although Follows & Bate of Manchester were bought by Qualcast in 1938, the Folbate brand name was used until 1966. The Folbate range saw few changes and the 1955 catalogue listed 10 and 12 in. cut J1 side-wheel mowers with cast iron wheels and wooden handle and the 12 in. cut F1 side-wheel mower with solid rubber tyres. The Falcon was a roller mower with a 12 in., six-bladed cylinder which, complete with grassbox, cost £5 plus 18s 3d purchase tax.

Suffolk Iron Foundry made the Super Clipper side-wheel mower, also the 10 in. Swift and 12 in. Super Swift roller mowers in the immediate

Fig 5.3   *Green's Clipper mower cost £1 5s 0d in 1939.*

Fig 5.4   *The Qualcast Panther cost £2 12s 3d. in 1938 and twelve years later it had increased to £7 2s 6d.*

post-war period. When the Stowmarket company joined the Qualcast Group in 1958 the Suffolk Swift and Super Swift were still made but the 10 in. and 12 in. Suffolk Viceroy side-wheel mowers had replaced the Super Clipper. The 10 in. Viceroy cost £3 19s 6d and the 12 in. Super Swift was £7 7s 0d.

By the late 1950s the Qualcast hand mower range was reduced to the tubular steel handled B1 and E1 side-wheel machines and the 12 in. cut Panther, which by this time cost £8 15s 9d. Production of the Suffolk Swift, Super Swift and Mark II Viceroy mowers continued into the 1970s and although the Folbate name had disappeared by 1966 the Falcon roller mower remained as the Suffolk Falcon until 1972.

The Qualcast Panther gave way to the Super Panther in the early 1960s and the E1 side-wheel mower was discontinued in 1967 after a run of almost 30 years. Qualcast mower design moved forward in 1969 with the launch of the 12 in. cut Superlite Panther. Diecast aluminium side plates and tubular steel handles with plastic grips

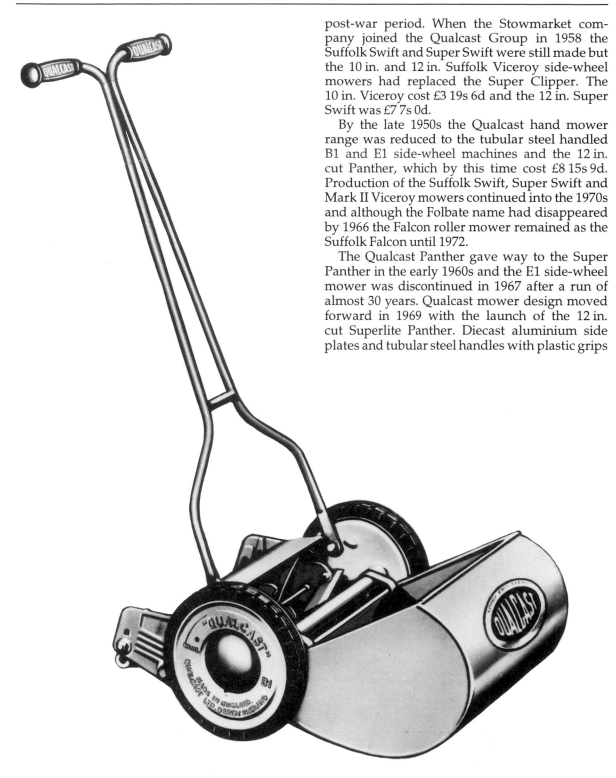

Plate 5.6   The B 1 side-wheel mower with solid rubber tyres and tubular steel handle cost £5 19s 6d with grassbox, when it was added to the Qualcast range in 1949.

**Plate 5.7   The Super Clipper side-wheel mower
was made by Suffolk Iron Foundry in the early
1950s.**

reduced the weight of the new model, which
had dirt seals on the bearings, caps over the oil
holes, an enclosed chain drive and twin nylon
'click' blade adjusters. The Superlite Panther cost
£8 19s 6d and carried a ten year guarantee.

The growing popularity of motor mowers
hastened the decline of hand machines and by
1973 all Atco and Suffolk models were self-
propelled with a petrol engine or an electric
motor. The Qualcast Superlite Panther and the B1
De luxe side-wheel were the only hand-propelled
models made by the group now trading as
Suffolk Lawnmowers.

The B1 side-wheel mower, first made in 1950,
was replaced by the 12 in. Qualcast Q7 in 1973.
The Qualcast Q30 side-wheel mower cost £29.95

Plate 5.8   The Suffolk Swift had a cast iron chassis, five-blade cylinder on self-aligning ball bearings and dirt-proof chain case. Complete with grassbox it cost £5 19s 6d in 1957.

when it superseded the Q7 in 1980. It had five self-sharpening steel blades, self-lubricating bearings and rubber-tyred wheels threw the clippings rearwards into a grass collector.

Henry Webb & Co. of Birmingham made their first hand mowers in 1928. The company's main products were spring forks and hub brakes for motor cycles, bicycle components and roller skates but by the late 1940s lawn mowers had become Webb's number one product. Known as the De luxe, the first Webb mower had an enclosed drive from the roller to the cutting cylinder and unlike any other mower at the time the side frames were pressed steel. The Webb design was half the weight of its competitors and less liable to breakage as all mowers up to then were made of cast iron.

A new factory was built in 1929 and the 12 in. Wasp, which cost £1 19s 6d, was added to the original 10 and 12 in. cut De luxe models. The 10 in. Whippet and high specification 12 and 14 in. Windsor hand mowers came later. Webbs also made the chassis for motor mowers made by the Enfield Motor Cycle Co.

Webbs reintroduced the pre-war models in 1947 with improved versions of the Whippet, Wasp and Witch remaining in production into the 1970s. The 10 in. cut Whippet was a ladies' lightweight mower weighing only 38 lb and the Wasp was a more robust machine with a 12 in., six-bladed cylinder. The 12 in. cut Witch with an eight-bladed cylinder for fine turf and twin rear roller giving differential and free wheel action, described in the 1969 Webb catalogue as the

Fig 5.5   The Folbate Falcon cost £2 8s 6d when Follows & Bate were acquired by the Qualcast group in 1938 and it was still being made by Suffolk Lawnmowers 30 years later.

queen of hand mowers, it cost £21 14s 0d.

H.C. Webb became part of Wolseley Hughes group in 1963 but continued to trade as a separate company until 1973 when the two garden equipment manufacturers combined to form Wolseley Webb. The Whippet was discontinued before Wolseley Webb became part of the Qualcast group in 1984. The Wasp and Witch hand mowers remained in production until 1988 when they cost £114.95 and £144.95 respectively.

Suffolk Lawnmowers were still making more than half a million hand mowers in the early 1970s but small and inexpensive electric models had reduced sales of the Suffolk Panther 30, 30 DL and 35 to about 10 per cent of that figure by the mid 1980s. Sales literature claimed that the Panther 30 DL 12 in. cut roller mower with a pivoting handle and a large diameter rear roller left a beautifully striped lawn. The Panther was first made in 1932 and the name survives in 1995 in the shape of the 30 cm Panther 30S and the side-wheel Panther 30 with a rear grass collector.

Plate 5.9   A steel roller, diecast aluminium side plates and a totally enclosed chain drive were features of the Qualcast Superlite Panther.

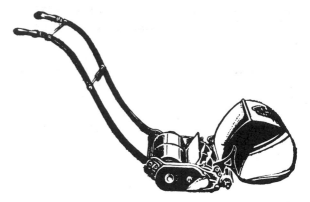

Fig 5.7   The Ransomes Atlas hand mower was made in the late 1930s.

Fig 5.6   The Webb Whippet cost £12 14s 0d in 1969 and the Witch, described as the queen of hand mowers, was £21 14s 0d.

Suffolk Lawnmowers' range of hand mowers in the early 1980s consisted of the Atco Elite 14 and 16 in. side-wheel machines together with the Webb Wasp and Witch roller mowers in the early 1980s. The Wasp and Witch were discontinued in 1989 when the Super Clipper 14 and 16 in. cut side-wheel mowers with a front-mounted steel grassbox appeared for the first time in the Atco catalogue.

Ransomes have traded under various styles in their two hundred year history. It was J.R. and A. Ransome who started making the Budding gear drive lawn mower in 1833. Ransomes & May set up a partnership in 1846 and this gave way to Ransomes & Sims in 1852. The Automaton roller mower with cutting widths from 10 to 18 in. was introduced in 1866 and another partner joined in 1869 to form Ransomes, Sims &

Head. Following a brief spell as Ransomes, Head & Jefferies from 1881, a private limited company trading as Ransomes, Sims & Jefferies was established in 1884. The Company went public in 1911.

Ransomes domestic lawn mower catalogue for 1938 included side-wheel and roller mowers. The Ace, Cub and Leo side-wheel models with prices starting at £2 1s 3d were described as admirable machines for small lawns. The Atlas and Anglia roller drive mowers, which cost from £3 8s 9d complete with grassbox, and the Ajax introduced in 1933, were said to be best suited to medium-sized areas of grass. Except for Ransomes gang mowers used to cut airfields, lawn mower production came to an almost immediate stop in 1939 and the factory was used to make armaments and components for aircraft.

The side-wheel Ripper was one of the first new models made at Ipswich after the war years. Introduced in 1946, the 14 in. cut Mark I Ripper was designed to cut grass up to 7 in. long. It was replaced by an improved Mark II in 1963 which remained in production until 1974.

The Ajax, Ariel, Astral and Certes roller mowers and the Mark I Ripper were shown in Ransomes hand mower catalogue for 1950. Mark I and Mark II Ajax mowers were made between 1933 and 1939. Production restarted in 1946 with the Mark III followed by the Mark IV and Mark V which was discontinued in 1972. Ten years later the Conquest, Ripper, Ajax, Ascot and Certes were being made at Ipswich. The Conquest was a sophisticated-looking 12 in. cut side-wheel mower with a roller chain driven six-blade cylinder under a streamlined sheet metal cover. The Conquest, made from 1959 to 1965, cost £7 11s 0d complete with a canvas grass catcher.

The much-improved Mark V Ajax was a lightweight easy-to-use mower. Ransomes' publicity material suggested that buyers would be proud to own an Ajax and even more proud of the velvety two-tone finish it would leave on their lawns. The 14 in. Ripper side-wheel mower and the long-established Certes with a 16 in., ten-blade cylinder for cutting fine turf completed the 1960s range of Ransomes' hand mowers. Ransomes discontinued their domestic mower range in the mid 1970s and expanded their interests in professional grass-care machinery.

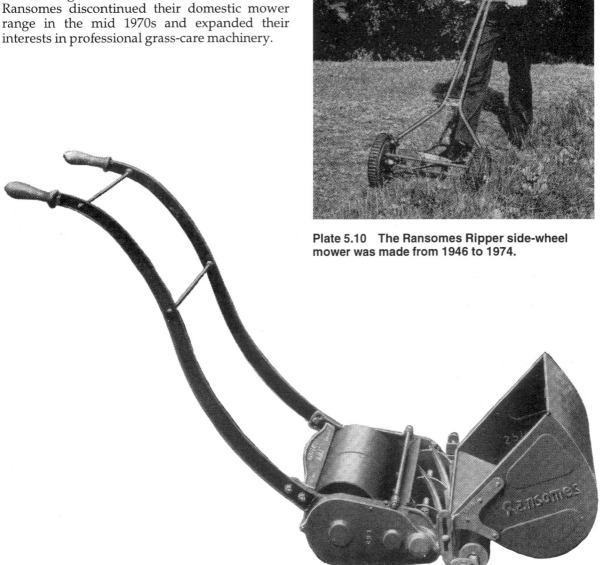

Plate 5.10   The Ransomes Ripper side-wheel mower was made from 1946 to 1974.

Plate 5.11   The Mark I Ransomes Ajax hand mower was made in 1933.

Plate 5.12   Ransomes Conquest was a disguised side-wheel mower.

MOTOR MOWERS

Lawn mowers were propelled by horses, ponies and steam engines in the late 1800s. A steam-powered, pedestrian-controlled lawn mower was made in 1893 and Alexander Shanks introduced a ride-on, steam-driven mower in 1900. Ransomes, Sims & Jefferies of Ipswich made the first commercial petrol-engined lawn mower in 1902 and Thomas Green introduced a 24 in. motor mower in 1903.

H.C. Webb launched a mains electric mower in 1947 and twelve years later this company exhibited a new battery mower at the 1959 Chelsea Flower Show. Not for the first time, mower design turned full circle in the 1960s with a new generation of mains electric cylinder and rotary mowers for the smaller garden. The next step, already well into the development stage, will see the return to battery mowers with rechargeable dry cells already in everyday use with cordless electric drills and other small power tools.

Charles H. Pugh Ltd of Birmingham pioneered the domestic motor mower for private gardeners in 1921. Earlier motor mowers were large and

Plate 5.13   This late 1940s Atco motor mower had a two-stroke engine with a kick start.

cumbersome and better suited to sportsgrounds and other large areas of grass. Atco's Managing Director Mr George Bull used a donkey to pull his cylinder mower but when one donkey died he decided to commission his works engineer to install an engine on his donkey mower. The project was a success and the mower was manufactured under the brand name of Atco, derived from the first and last letters of the Atlas Chain Company, also owned by Charles H. Pugh Ltd.

Arundel & Coulthard of Preston, established in 1815, were making Presto hand-pushed and petrol-engined cylinder mowers including the

14 in. cut Presto mower with an air-cooled two-stroke petrol engine which cost 18 guineas in 1938. Presto mower production resumed soon after the war in the mid 1940s and continued until the late 1960s when Qualcast purchased Arundel & Coulthard and closed the business.

Qualcast announced their first motor mower in 1948. The Qualcast 16 had a 98 cc Villiers two-stroke engine, 16 in. cylinder and cost £47 18s 7d complete with grassbox. The 16 in. cut, engine-driven Commando side-wheel mower for long grass was announced in 1953 and the 16 in. Royal Blade motor mower priced at £48 7s 4d was added in 1954.

Suffolk Iron Foundry introduced the famous Punch motor mower in 1954. Three years later the Suffolk motor mower catalogue included the 12 in. cut Auto Swift and Pony mowers, also standard, Super and Professional models of the Punch. Most of these mowers had dual drive, which allowed the user to disengage drive to the

Plate 5.14   The Qualcast 16 motor mower was introduced in 1948.

Fig 5.8   A 1937 advertisement for the Presto motor mower made by Arundel & Coulthard. Qualcast acquired the company in the late 1960s and closed it down.

Plate 5.15   The first Suffolk Punch motor mowers were made by Suffolk Iron Foundry at Stowmarket in 1954.

roller and push the machine when cutting in confined spaces or awkward corners.

A 50 cc Suffolk two-stroke engine and dual drive was used for the Pony and the Punch had a centrifugal clutch, self-aligning ball bearings and enclosed drive to the 14 in. cylinder. Dual drive was added to this model in 1956. A sales leaflet suggested that the four-stroke engine with a recoil starter on the 14 in. Punch used $\frac{1}{4}$ pint of fuel to cut 500 square yards of grass in less than 20 minutes. The 17 in. dual-drive Super Punch was introduced in 1956 and the Super Punch Professional with a 17 in., ten-bladed cylinder was used for fine turf.

The Squire and Squire Corporation four-stroke-engined side-wheel mowers with a centrifugal clutch and rubber-tyred wheels completed the 1957 list of Suffolk motor mowers. The Squire Corporation was renamed the Corporation in 1960. It had a heavy-duty bottom blade and a safety clutch to protect the

drive to the 19 in. cylinder if it was temporarily overloaded.

Qualcast introduced a motorised version of the Panther roller mower in 1954. The Powered Panther with a 34 cc J.A.P two-stroke engine used about one pint of fuel per hour and cost £24 19s 6d. It was not self-propelled but publicity material pointed out that the power-driven cutting cylinder lightened the gardener's load. A foot pedal was used to engage drive to the cylinder and it was disengaged automatically when the user released the handles.

The Qualcast Royal Blade De luxe, Powered Panther and the Commando were still being made in 1957; the Commando was discontinued in 1960. Qualcast and Suffolk Iron Foundry retained their own models of motor mower when the Stowmarket company joined the Qualcast group in 1958. Early co-operation was evident with the Qualcast Commodore launched in 1959. Unlike the two-stroke-engined

Royal Blade De luxe, the new 14 in. Commodore with dual clutch control and recoil starter had a Qualcast–Suffolk 75 G four-stroke engine.

The 12 in. cut Suffolk Colt with a four-stroke engine, automatic clutch and dual drive replaced the Pony in 1960. Suffolk Iron Foundry also added the new Microset cylinder adjustment to the 14 in. Punch and announced the Carry-Mow mower transporter in the same year.

The Carry-Mow was a wheeled cradle placed under the rear roller, which made it easier to push the Suffolk Punch, and other mowers, up and down steps or across gravel paths. The Carry-Mow cost £3 9s 6d and additional fittings priced at £2 4s 0d converted it to a sack holder or household bin carrier.

The Suffolk Iron Foundry catalogue for 1965 included the new Mark II Corporation side-wheel mower and various models of the Colt,

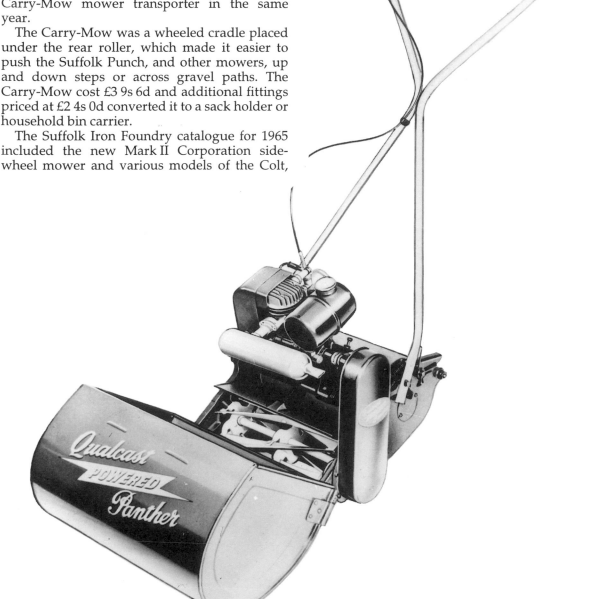

Plate 5.16   A foot pedal was used to engage the drive to the Qualcast Powered Panther's cutting mechanism.

Punch and Commodore. The 12 in. cut Super
Colt introduced in 1968 was described as
streamlined in construction with many special
features including an automatic clutch, dual
drive operated from the handles, unique bottom
blade assembly and specially hardened cutting
cylinder blades.

    Four models of Suffolk Punch were in the 1969
price list but the Commodore was no more.
The Mark II Corporation was £29 19s 6d but the

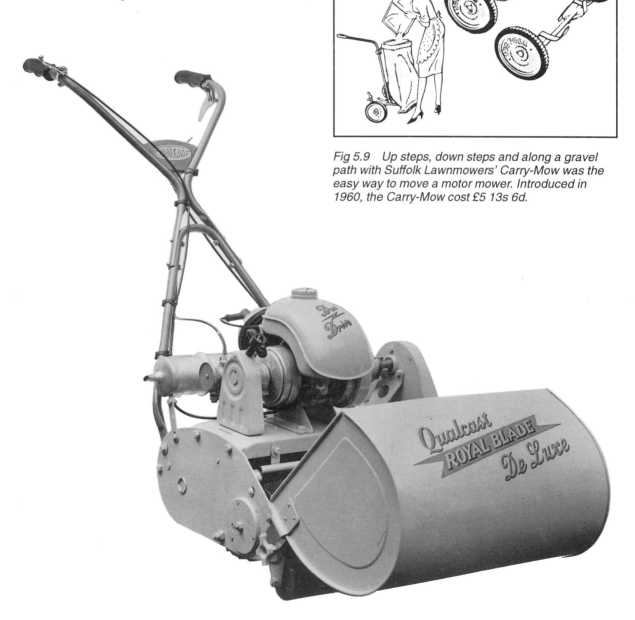

Fig 5.9   Up steps, down steps and along a gravel
path with Suffolk Lawnmowers' Carry-Mow was the
easy way to move a motor mower. Introduced in
1960, the Carry-Mow cost £5 13s 6d.

**Plate 5.17   The Qualcast Royal Blade De luxe 14 in. cut motor mower with recoil starter and a dual
clutch control cost 39½ guineas in 1957.**

Squire had given way to the self-propelled Suffolk 16 with a Suffolk 75G four-stroke engine and centrifugal clutch.

Atco had joined the Qualcast Group in 1962 and by the late 1960s there were twelve different Atco motor mowers from a 12 in. domestic model to a 34 in. professional machine.

The range of Qualcast and Suffolk motor mowers was gradually reduced after Atco joined the Qualcast group in 1962. However, by the late 1960s there were twelve models from a 12 in. domestic mower to a 36 in. professional machine carrying the Atco name. British currency went decimal in 1970 and Value Added Tax at 8% was included in the price of £55.95 for a 14 in. Suffolk Super Punch.

Further shrinkage left four petrol-engined mowers, the 12 in. Super Colt, 14 in. Super Punch, and two models of the 17 in. cut Super Punch, on Birmid Qualcast's 1975 price list.

Qualcast Panthers were either mains or battery electric mowers and the Atco domestic range was reduced to 12, 14, 17 and 20 in. wide cutting cylinders with separate clutch levers for the roller drive and cutting mechanism. Although improved, the same four mowers were still made in 1980 but the prices were very different. The 12 in. Atco was £36.25 in 1970 but ten years later the influence of inflation and 15% VAT had

Plate 5.18   The Atco Toughcut side-wheel motor mower.

increased the price to £132. The Birmid Qualcast Group sold more than 100,000 petrol-engined mowers in 1975 but the preference for electric mowers had reduced this figure to less than 40,000 by 1985.

Atco introduced the B14, B17 and B20 De luxe motor mowers and an inexpensive 12 in. model for small gardens in 1969. The model number indicated the width of cut, all four had a 98 cc four-stroke engine with electronic ignition and side wheels could be used instead of the front roller for long grass. The Atco Commodore, with

Plate 5.19   The Atco B14, B17 and B20 De luxe motor mowers were introduced in 1969. The 14 in. model cost £39 19s 0d.

Plate 5.20   Launched in 1984, the Atco Commodore revived a model name used by Qualcast in the 1960s.

a new 114 cc aluminium engine and single lever safety control on the handlebars, replaced the De luxe mowers in 1983. Apart from the 12 in. model, discontinued in 1990, the three Commodores with plastic grassboxes and optional electric starting are listed along with the Ensign, Royale and Club in the 1995 Atco catalogue. The Ensign is a less expensive mower for the medium-sized garden, the Royale with a ride-on seat is suitable for large lawns and the Club, first made in 1983, has a ten-bladed cylinder for fine turf.

Ransomes' first petrol-engined lawn mower was made in 1902 and by the mid 1930s the Ipswich company were making at least nine different motor mowers from the 14 in. Midget, with a 1 hp air-cooled two-stroke engine, to the

massive 11 hp, 42 in. cut ride-on machine, which weighed 22 cwt. The Midget cost £22 17s 6d and the 42 in. cut mower was £325 but five per cent discount was allowed for cash.

Many of the pre-war models were continued after the war years and by the early 1950s ten two- and four-stroke motor mowers were listed in Ransomes' catalogues. The 12 and 14 in. cut self-propelled Minor and the 18 in. Auto-Certes had Villiers 98 cc two-stroke engines and a centrifugal clutch. The Auto-Certes for bowling greens and croquet lawns was hand-propelled; the catalogue stated that 15 cuts per inch were achieved with the ten-blade cylinder at 3 mph. The 147 cc Villiers two-stroke-engined Ransomes 16 was recommended for private lawns and tennis courts. It had a centrifugal

Plate 5.21 The six-bladed cylinder and rear roll on the Ransomes Minor motor mower were roller-chain driven by a Villiers two-stroke engine.

clutch with separate dog clutches on the drive roller and 16 in. cutting cylinder.

The Gazelle, with a light alloy die cast frame, 18 in. cylinder and a 98 cc Villiers two-stroke engine, was the only Ransomes powered side-wheel mower in the early 1950s. The 20 in. cut Antelope replaced the Gazelle in 1956; eight years later the Mark 4 Antelope with a B.S.A. four-stroke engine and a grass catcher cost £53. The Antelope was discontinued in 1993.

Ransomes Mercury, Marquis and Sprite were, according to a 1960 sales leaflet, the mowers for the discriminating gardener. A 34 cc J.A.P two-stroke engine powered the 14 in. cut Sprite and with fingertip dual control from the handlebars

it was said to cut 500 square yards of grass for less than 1d. The Sprite, which cost £32 17s 3d, was discontinued in 1964.

The 16 in. cut, 75 cc Villiers four-stroke-engined Mercury was made between 1958 and 1964. A de luxe version of the 18 in. cut Marquis appeared in 1964. Dual control was standard on the Mercury and the Marquis; both were claimed to cost less than 3d an hour to run.

The Ransomes Fourteen, announced in 1964, was described as a quiet-running quality mower, packed with power and made to last—a slimline machine, light and manoeuvrable enough for any member of the family to use. The Fourteen was made for six years. The petrol-engined version with a B.S.A. 65 cc four-stroke, later replaced with a 65 cc Villiers, cost £42 10s 0d and the 12 volt battery electric model was £52 10s 0d.

Ransomes withdrew from the domestic market in the mid 1970s to concentrate on professional grass-care equipment including the Marquis, Matador, Mastiff and the Twenty Four motor mowers. Although the last named was discontinued in 1981, the others are still being made in 1995, some 160 years after J.R. and A. Ransomes were licensed to manufacture Edward Budding's gear drive mower.

H.C. Webb were making a range of motor mowers before they became part of the Wolseley Hughes group in 1963. New Webb 12 and 16 in. cut motor mowers, described as very fine

Plate 5.23   The Ransomes Antelope side-wheel mower, which replaced the Gazelle in 1956, was made until 1993.

Plate 5.22   Ransomes Marquis Mark 4 motor mower with a B.S.A. 119 cc four-stroke engine.

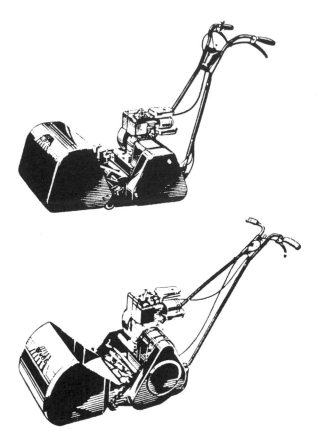

Fig 5.10   *Some of the 1969 Webb electric and battery mowers.*

circle that made a London taxi look like the Queen Mary. The ride-on mower, which cost £232, had ribbed rubber-covered rollers, lever-operated clutch, double vee-belt transmission and footbrake.

The Webb AB series, later called the Wizard, was introduced in the early 1970s with the option of 18 and 24 in. pedestrian models and a 24 in. ride-on. Another new mower with a 3 hp Briggs & Stratton engine and a 21 in. six-bladed cylinder on self-aligning ball bearings appeared in 1981. A completely new range of Webb motor mowers followed in 1982. There were three models with Briggs & Stratton engines, the 14 and 18 in. cut mowers had a 2 hp power unit and the 5 hp, 24 in. cut model was made in pedestrian-controlled and ride-on format.

Wolseley Webb joined Suffolk Lawnmowers in 1984 and except for the 24 in. Webb motor mower it was all change in 1985 when new designs brought Webb petrol models into line with the Suffolk Punch. Webb motor mowers have not been made since 1988, except for a brief appearance of the Webb Diplomat in 1990 and 1991.

and sophisticated machines, were introduced in 1958. The 16 in. model had a detachable power unit and clutch, which could be used to drive other garden appliances including a hedge trimmer and hand tiller with a flexible shaft.

Lower-priced 14 and 18 in. cut Webb petrol mowers with Briggs & Stratton 2 hp engines and dual drive were introduced in 1961 and a 24 in. ride-on was added to the range in 1963.

H.C. Webb Ltd, now part of Wolseley Hughes, became Webb Lawnmowers Ltd in the late 1960s and new models for medium-sized and large lawns were announced in 1969. A 24 in. cut ride-on with a trailed bucket seat carried on a roller was described in sales literature as fast, wide and handsome. It suggested that new owners could leap aboard and ride ahead of their cutting problems with a mower that cared for the lawn and its owner. The close-coupled seat was claimed to give the machine a turning

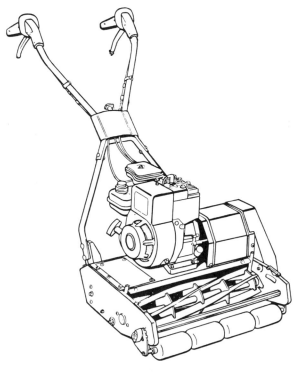

Fig 5.11   *The Wolseley Webb Wizard motor mower.*

Plate 5.24   The Haytermower could be used with a 30 in. cylinder unit or a rotary mower deck.

The original Hayter cylinder mower, introduced in the late 1950s, came about after Douglas Hayter realised that one of his competitors, well known for their lawn mowers, was moving into the rotary mower market. The self-propelled Haytermower with interchangeable cylinder and rotary cutting units was designed and built in a matter of weeks.

Two models of the self-propelled Ambassador, launched in 1967, were the first purpose-built Hayter cylinder mowers. The 16 in. Ambassador had a B.S.A. 119 cc engine and a 2¼ hp Briggs & Stratton power unit was used on the 20 in. model. A professional version of the 20 in. Ambassador with a Villiers four-stroke engine and ten-bladed cylinder was introduced in 1969. An improved Ambassador 2 with a five-bladed, 20 in. cylinder and 3 hp Briggs &

Stratton engine was announced in 1973 and the ten-bladed Ambassador Super 2 was added in 1979.

The Hayter 30 Condor, introduced in the late 1960s, with a 6 hp MAG engine and three speed gearbox, was designed for parks and sports grounds. The Condor's 30 in. cutting cylinder, like that on the earlier Haytermower, could be swapped with a 30 in. twin-bladed rotary mower deck.

The Hayter Senator 30 replaced the Condor in 1980, it had an 8 hp Kohler engine and a hydrostatic drive unit with an infinitely variable speed from 0 to 5 mph in both directions. A differential unit was built into the split rear rollers and cylinder speed varied with forward speed.

Two versions of the Ambassador 3 with five- and ten-bladed 20 in. cylinders and a Briggs &

Plate 5.25   Introduced in 1967, the Ambassador
was the first cylinder mower made by Hayters of
Spellbrok for the home gardener.

Plate 5.26   The Merry Lawn Mower could be
attached to the Merry Tiller in a minute without
using any tools.

Stratton Quantum engine replaced the previous
model in 1992. Two models of the Ambassador 3
and the Senator 30 were included in the Hayter
catalogue for 1994.

Cylinder mower attachments were made for
some 1950s garden cultivators including the
Trusty Earthquake, Merry Tiller, Howard Bantam, Barford Atom and Allen Motor Scythe. The
Merry Lawn Mower was attached to the Merry
Tiller rotor shaft housing and forward travel
was controlled with the cultivator clutch. A
vee-belt driven 16 in. Webb mower unit with a
separate clutch made it possible to push the
machine around flower beds and reach into
awkward corners.

The Trusty Mowmotor made by Tractors
(London) Ltd was a small power unit for hand-
propelled cylinder mowers in the mid 1950s.
Advertising material described the Mowmotor

with a $\frac{1}{3}$ hp two-stroke engine as a sturdy,
dependable little job at a reasonable cost, which
would fit any ordinary mower and do two or
three hours' work on a pint of fuel. The weight of
the Mowmotor held two small diameter rollers,
chain-driven from the engine against the back
roller on the mower. A lever on the handlebars
was used to raise and lower the engine unit to
start or stop forward travel.

Electric lawn mowers can be traced back to the
1890s but the first commercial mains electric
mower was made by Ransomes in 1926. H.C.
Webb introduced a mains electric mower in 1947
but the real revolution in powered lawn mower
design got underway in the late 1950s. Webbs
created something of a sensation at the 1959
Chelsea Flower Show when they launched a 12
volt battery mower with the added gimmick of
radio control. By the mid 1960s Webbs were
manufacturing two-speed 12 and 14 in. cut battery mowers with a built-in $1\frac{1}{2}$ amp charger and a
de luxe 14 in. model with a 3 amp charger.

The Ransomes Electra, the first mains electric
lawn mower to be sold in quantity, appeared in
1926. The 14 in. cut model, with a $\frac{1}{2}$ hp motor,
weighed $1\frac{1}{2}$ cwt and was said to cut half an acre in
an hour and consumed $\frac{1}{2}$ a unit of electricity.
The Electra was also made with 16 and 20 in.
cylinders and a 30 in. model could be made to

Plate 5.27 A sales leaflet suggested that the Trusty Mowmotor attachment would put an end to the labour and sweat of pushing a mower for hours on end every time there was a blade of grass to cut.

order. A 1933 price list included the 14 in. mower at £27 10s 0d and the 20 in. model was £57 10s 0d. A Ransomes catalogue explained that in view of the modern practice to use electricity wherever possible to drive labour-saving devices, it was in the trend of events that this form of power should be adapted to the lawn mower. The chief advantages of electric mowers were said to be cleanliness, silent operation, economical running and easy manipulation.

Mains electricity was not widely available in country districts in the 1930s and this fact together with the outbreak of war brought production of the Electra to an end. In common with other mower manufacturers, Ransomes added mains electric mowers including the Minor, Mercury and Majestic to their product

range in the mid 1950s. Battery mowers became popular in the early 1960s and the Ipswich company made a battery electric version of the Ransomes Fourteen motor mower from 1964 to 1969.

H.C. Webb were making mains electric mowers in 1948 and continued to do so for the next 20 years. A 1969 catalogue included battery and mains electric mowers with 12 and 14 in. cutting cylinders, which cost less than a penny an hour to run. Four years later Webbs introduced low voltage mains electric models with a transformer to reduce the power to 110 volts.

Wolseley Webb had progressed to 12 and 14 in. cut mains electric and 12 in. battery versions of the Wizard motor mower by the late 1970s. The hand-pushed 12 in. cut mains model cost £33.73 and the most expensive 12 in. battery model was £82.22.

The Qualcast Panther took on a new form in 1962 as the 12 in. cut Super Panther Electric with a 0.4 hp mains electric 220/250 volt motor, 75 ft of cable and dual drive for cutting in confined spaces. A swinging arm on the handlebars kept the cable away from the cutting cylinder while the self-propelled mower travelled back and forth across a lawn. The three-speed, 14 in. cut

Super Panther Electric, which cost £29 19s 6d, was said to allow the user to walk at a pace to suit the length of the grass and available energy was added in 1965.

A battery-powered 12 in. cut Super Panther was announced for the 1967 season, complete with a built-in 12 volt battery charger; the new model cost £29 19s 6d. It was an immediate success and although about 10,000 were produced for its first season, demand for

Fig 5.12 Qualcast added a mains electric version of the Super Panther in 1962.

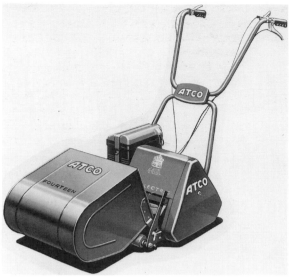

the battery electric Super Panther outstripped supply. Twice as many were made in 1968 and production continued for another eleven years.

The 12 in. cut Qualcast Concorde electric cylinder mower was launched in 1971. Complete with 50 ft of cable it cost £15.75. It was a huge success with sales not far short of half a million in the first two years. This led to the introduction of the 14 in. cut Astronaut with 75 ft of cable in 1974 but it was not popular and was discontinued in 1979. The Qualcast Concorde has been made in huge numbers in various forms including the 12 in. Concorde E30 for short or long grass introduced in 1980, the Concorde RE 30 DL with a rear grassbox and the 14 in. cut Concorde

Plate 5.28 **The Atco Fourteen was added to the range of 12, 17 and 20 in. cut battery electric mowers made by Charles H. Pugh Ltd in 1966.**

Plate 5.29 **A mid 1930s Rotoscythe, the world's first rotary mower.**

Plate 5.30 A two-stroke engine provides the power for the 16 in. cut Rotoscythe 16, which cost £38 15s 0d plus £10 14s 0d purchase tax in 1949.

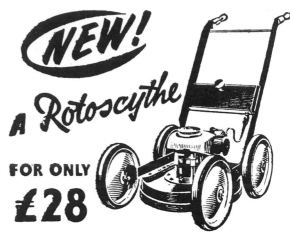

Fig 5.13   The 18 in. cut Rotoscythe County mower was introduced in 1950.

RE 35 DL. Still going strong in 1995, the Concorde range consists of the 12 and 14 in. E30 and E35 with front grassbox also the 10, 12 and 14 in. XR models with a rear grassbox.

Atco were making 14 and 17 in. cut battery electric mowers in 1964 and a 12 in. model was added in 1965. All three had a 12 volt battery with a built-in charger and two-speed drive. Battery electric mowers were popular for a period in the late 1960s but sales fell away and Atco dropped them from their product range in 1975.

Atco De luxe mains electric cylinder mowers with an improved drive mechanism and pram-type handlebar appeared in 1977. The 12 in. model with a lightweight polypropylene grassbox cost £37.29. Mains electric versions of the Atco Commodore and Royale motor mowers replaced the De luxe mowers in 1986 and are still produced in 1995.

ROTARY LAWN MOWERS

The world's first rotary mower, the Rotoscythe, was made by Power Specialities Ltd at Maidenhead in 1933. The original machine was the result of an experiment by David Cockburn to make a rotary hedge trimmer combined with a domestic vacuum cleaner to cut his hedges and collect the trimmings. It was not a success and as he dragged his invention across a lawn

with the motor still running Cockburn noticed that as well as cutting a strip of grass it had also picked up the clippings. A prototype was made and after patents were taken out to protect the vacuum collection system and rear-attached grassbox, the Rotoscythe went into production.

It was made until the outbreak of war and soon after production started up again in the mid 1940s, Power Specialities introduced the new Rotoscythe Model 16. Company advertising pointed out that the new mower had 'a very wide range of usefulness, was of compact form and handy in operation'. A two-stroke engine powered the cutting rotor and forward travel but the clutch could be disengaged when it was

Plate 5.31   The Loyd Motor Sickle, with a 98 cc Villiers two-stroke engine, exhibited at the 1950 Smithfield Show, cost £39 15s 0d.

necessary to push the Rotoscythe into awkward corners. The 16 in. rotor with impeller blades to blow the clippings into a rear-mounted grassbox was bolted to the engine's vertical crankshaft. A set of three moss removal pegs, which cost 6s 1d, could be fitted to the rotor in place of the cutting blades.

A more robust 18 in. cut County Rotoscythe designed for the market gardener and smallholder cost £28 when it was launched in 1950. The County did not have a grassbox and

returned the trimmings to the ground. This followed the trend in mower design to leave the trimmings on the surface but this fashion did not last and within a few years a grassbox was standard equipment on most rotary mowers.

The Rotoscythe patents were extended beyond the normal term because of the war but they eventually expired in 1952 and it was not long before other companies were making similar rotary mowers. Power Utilities were acquired by J.E. Shay Ltd of Basingstoke in 1952. They continued making the Rotoscythe and used the two-stroke mower engine for the Shay Roto-gardener cultivator in 1954.

The hand-propelled Loyd Motor Sickle made by Vivian Loyd & Co. at Camberley in Surrey appeared in the late 1940s. The Motor Sickle was an 18 in. cut rotary mower with a 98 cc Villiers two-stroke engine and vee-belt drive to the cutter head. Although it was mainly used to cut rough grass and scrub, the mower disc could be replaced with a 14 in. circular saw blade,

claimed to cut through trees and saplings up to 8 in. diameter. Loyd's publicity material stated that the Motor Sickle used no more than a gallon of petrol in an eight-hour day. No mechanical knowledge was required by the operator who, it was said, could use it all day long without fatigue. It was also pointed out that blunt cutter blades could be sharpened in position or replaced in less than three minutes at a trifling cost.

One satisfied user wrote that the Loyd Motor Sickle not only cut coarse grass, garden produce, brambles and undergrowth but it also cut labour, time and costs, everything in fact except your taxes and your hair!

Electric rotary mowers similar to those now sold in their thousands were made in the late 1940s. The Ladybird mains electric rotary mower with a single handle, front roller and high-speed blades minced the grass so finely that raking and sweeping was unnecessary.

Plate 5.32   The 24in. cut Hayter Motor Scythe was introduced in 1947 and remained in production for 26 years.

Plate 5.33   This neatly dressed gentleman was able to cut close to his fruit trees with this 1960 Hayter Motor Scythe, and unlike earlier models a guard protects the user's toes from the high-speed blades.

Advertisements at the time pointed out that the Ladybird, which cost 16 guineas (£16 16s 0d) in 1950, was as easy and economical to use as a domestic vacuum cleaner. It was suitable for AC or DC mains supply, weighed a mere 16 lbs and was said to be so manoeuvrable that a child could use the Ladybird in complete comfort and safety.

Douglas Hayter, who started as a building contractor in the 1930s, achieved his ambition to establish a manufacturing enterprise. He started his business after making a rotary mower from various bits and pieces at hand. Much to his surprise it worked quite well but a complete lack of guards made it a rather lethal creation with stones and debris flying in all directions.

Plate 5.34   The slogan said 'Sooner or later you will buy a Hayter'. The first Hayterette mowers were made in 1957 and cost £36 delivered in England and Wales. Easy payment terms were available. More than a quarter of a million have been made in the last 35 years.

The first production models of the Hayter Motor Scythe were made at Spellbrook in 1947. They were simple hand-pushed machines with vee-belt drive to the 24 in. cutting rotor from whatever suitable engine was available at the time. When supplies became more readily available a four-stroke Villiers horizontal crankshaft engine was fitted to the Hayter Motor Scythe. The near-side rear wheel was inset to allow grass at the side of the lawn to be cut without the mower dropping off the edge. Sales literature claimed that coarse, tough grass difficult to scythe by hand and impossible to cut with any other type of mower could be mastered with the greatest of ease with a Hayter Motor Scythe.

In common with most 1950s garden tractors and cultivators, other attachments including a fruit tree sprayer and a generator could be used with the Motor Scythe. The unit, which cost £21 10s 0d, had a vee-belt driven pump, with a suction hose and hand lance. The belt-driven 400 watt, 110 volt DC generator, which cost £19 15s 0d, was bolted to the Motor Scythe (Plate 6.17). Recommended Tarpen equipment included hedge trimmers, Hedgemaster, Grassmaster and a $\frac{1}{2}$ in. drill chuck.

The Spellbrook company turned their attention to tractor-drawn mowers. A 5 ft cut Hayter Orchard Mower appeared in 1950 and the 6 ft power take-off driven Haytermower was added in 1953. A 26 in. cut Hayter rotary mower and the 6 ft 6 in. Highway Mower for professional and highway authority use were launched in 1956.

Hayters entered the domestic rotary mower market in 1957 with the Hayterette and although there have been many improvements over the years it remains in production to the present day with total sales now well over 250,000 machines. A variety of colour schemes have been used on the Hayterette since the original red and silver livery, including the distinctive hammer green in the 1960s, followed by red and green and the dark green paintwork of the 1990s.

The Hayterette and other rotary mowers that returned the clippings to the ground did not leave the traditional striped picture-book lawn. The Spellbrook company introduced two versions of the Haytermower in the late 1950s to meet the demand for a striped finish with a rotary mower. A self-propelled Haytermower designed for large areas of grass had an output of up to half an acre per hour and a hand-pushed

Plate 5.35   A 26 in. rotary mower deck on the Haytermower, introduced in 1959, was interchangeable with a 30 in. cylinder mower unit.

model with a grassbox under the handlebars was suitable for medium-size gardens. The 26 in. cut rotary deck on the self-propelled model with split roller drive was interchangeable with a 30 in. cylinder mower attachment. It had three gears with a top speed of $4\frac{3}{4}$ mph and the rotary cutter height was adjustable from $1\frac{1}{2}$ to 6 in. A wheeled or roller-mounted trailed seat, a centrifugal self-priming pump, generator and a flexible drive for a Tarpen hedge trimmer were included in the list of accessories.

The 18 in. hand-propelled Haytermower used a vacuum system to pick up the clippings or leaves and the nearside wheels were inset to

Plate 5.36   The Bank Rider, launched in 1974, was the first Hayter ride-on rotary mower.

allow the mower to cut right up to the edge of the lawn. A deflector plate, used in place of the grassbox, was supplied for cutting areas of rough grass.

The factory at Spellbrook was enlarged in 1964 to provide extra production space for an increased range of hand-propelled, self-propelled and tractor-drawn rotary mowers.

A Briggs & Stratton four-stroke engine, with a recoil or a wind-up starter, was used on the 1967 Hayterette and a J.L.O. two-stroke engine was fitted to the professional model. The recoil-start Hayterette cost £37, just £1 more than the original machine ten years earlier. The safety of the user was also taken into account with an optional guard kit, which cost £2 5s 0d to comply with new regulations for powered mowers used in agriculture.

Hayters adopted various bird names for their

mowers in the early 1970s. The hand-propelled 12 in. Hawk and 19 in. Hawk Major rotary mowers were available with wheels or rollers. The self-propelled range included the Harrier, Osprey and Condor with cutting widths from 19 in. on the $3\frac{1}{2}$ hp Briggs & Stratton-engined Harrier to the 30 in. Condor with a 6 hp MAG engine. The hand-propelled Merlin, also with a $3\frac{1}{2}$ hp Briggs & Stratton engine and grassbox, was added in 1973 and like the roller-driven Harrier left the lawn with the much admired light and dark green stripes.

The Bank Rider, introduced in 1974, was the first Hayter ride-on rotary mower and as the name suggests it was designed to cut sloping grass surfaces including those at the side of public highways. It was one of the first mowing machines to have hydraulic motor drive to the cutting deck and the off-set operator's seat

Plate 5.37   The Hayter Frigate was made from 1976 to 1980.

could be swivelled so that it remained level
when driving across a sloping surface. About a
hundred Bank Riders were made and most of
them were sold in Lancashire.

The 36 in. Hayter Frigate rotary mower
replaced the Bank Rider in 1976 and the 53 in.
cut Eagle was added in 1978. Ride-on and
pedestrian-controlled versions of the Frigate,
designed for professional and public authority
use, were powered by a 7.3 hp Kohler four-
stroke petrol engine. Transmission was through

a transaxle unit with three forward gears and
reverse, differential and final drive. Interchangeable front-mounted rotary and cylinder mowing
decks were driven from the engine crankshaft
through a hand clutch and vee-belt. A drawbar
was provided for trailed equipment.

Three cutter heads were vee-belt driven from
a 16 hp Kohler engine on the self-propelled
Eagle rotary mower. A transaxle unit provided
three forward speeds, a creeper gear and reverse
with a top road speed of 8 mph. The Frigate was

discontinued in 1980 but the Eagle remained in production until 1983.

The 1979 Hayter price list included the pedestrian-controlled Frigate at £1,043, the ride-on model was £1,253, the Eagle cost £3,235 and the price of the 18 in. Hayterette had risen to £150.65p.

Changes at Spellbrook saw Hayters listed on the Stock Exchange in the early 1980s and soon after this Hayter plc were bought by F.H. Tomkins, a financial holding company who acquired the Murray Ohio Manufacturing Co. in 1988. Murray were the world's largest mower manufacturer at the time and before the take-over G.D. Mountfield were their UK distributors. The change of ownership resulted in Murray rotary mowers, including a 30 in. cut 10 hp rear-engined ride-on machine and an 18 hp garden tractor with an underslung mower deck, being added to the Hayter range.

The Hayter Hobby, with a strong plastic

Plate 5.38 The Hayter Hunter range, launched in 1987, included push and self-propelled 16, 18 and 21 in. cut machines. Electric starting was standard on 18 and 21 in. self-propelled models by 1989.

grassbox and a 3½ hp Briggs & Stratton engine to drive the 16 in. cutting rotor, appeared in 1982. Designed for the small- and medium-sized lawns, the Hobby had a split rear roller, claimed to make it easy to push and leave the banded finish so highly prized by keen gardeners. Mains electric and Tecumseh-engined self-propelled versions were added to the Hobby range in 1985.

Tomkins purchased Beaver Equipment Ltd, makers of grass machinery for golf courses, parks, amenity areas, etc. in 1987 and production of this machinery was transferred to Spellbrook in 1991.

Ransomes entered the rotary mower market in 1960 with the dual purpose Mark I Multimower. Designed to cut large grass areas, it had a J.A.P 288 cc four-stroke engine, two speed gearbox and interchangeable 27 in. rotary and 30 in. cylinder mower attachments. Mark II and Mark III Multimowers, also with J.A.P engines, followed in 1961 and 1969. The Multimower 2000, still with the optional cylinder and rotary mower deck, was announced in 1972. Further models with 4 hp MAG engines were made until the mid 1980s when Ransomes introduced a new range of professional rotary mowers and the Multimower was listed as a reel cutter with a 30 in. cylinder.

The Ransomes Cyclone 18 in. rotary made in the late 1950s with a 2 hp J.A.P two-stroke engine was described in a sales leaflet as the mower with a hummock disc. This was a convex disc under the two-bladed rotor, said to ride over uneven ground to protect the cutters and prevent long grass binding round the shaft and stalling the engine.

Two- and four-stroke versions of the 18 in. cut Ransomes Typhoon replaced the Cyclone in 1961. They cost £21 15s 0d and £26 15s 0d respectively and a magazine advertisement at the time suggested these were very low prices indeed. In spite of this advantage the Typhoon was withdrawn a few months after Ransomes introduced the 18 in. Typhoon Major with a Villiers 150 cc four-stroke engine in 1962. It was not a simple task to alter the cutting height of rotary mowers in the early days and the Typhoon was no different to most of its competitors. Spanners and sometimes a hammer were needed to alter the position of the wheels on the rotor housing.

The Cyclone and Typhoon mowers returned the clippings to the ground but the new Ransomes 18 in. Rotary, announced in 1965, had

Plate 5.39   Ransomes introduced the 18 in. cut Typhoon rotary mower in 1961.

Plate 5.40   This Ransomes 18 in. cut rotary mower with a grassbox was introduced in 1965.

a large metal grass collector and a single lever provided instant cutting height adjustment. Although the 18 in. Ransomes Rotary survived for only two years, the Typhoon Major remained in production until 1969.

Ransomes concentrated their efforts on professional rotary mowers during the 1970s and early 1980s but the acquisition of G.D. Mountfield at Maidenhead and Westwood Engineering at Plymouth in 1985 gave the Ipswich company a renewed interest in the domestic mower business.

G.D. Mountfield was established at Maidenhead in 1962 and one of the first products was a 3 hp four-stroke cultivator with a rotary mower attachment. The 18 in. cut Mountfield M3, Minor and Major with $3\frac{1}{2}$ hp four-stroke Aspera engines

in hand-pushed and self-propelled versions were made in the 1970s.

The Mountfield rotary mower catalogue for 1986 included hand- and self-propelled versions of the Empress, Emperor, Emblem and Mirage with Briggs & Stratton, Tecumseh and Suzuki petrol engines. The 15 in. cut hand-propelled Emblem was also made with 240 volt electric motor. The 21 in. cut Monarch was added in 1988 and the Laser range was launched in 1993.

A new 18 in. cut heavy duty, hand-propelled rotary based on a mower originally designed by Mountfield was launched by Ransomes in 1989. It has a 5 hp Mag Kubota engine mounted on a cast alloy deck with an optional rear-mounted grass collector.

Westwood Engineering were making a range

Plate 5.41   A modern version of the Ransomes 18 in. rotary mower appeared in 1988.

of garden tractors and pedestrian-controlled mowers when they became part of Ransomes, Sims and Jefferies in the mid 1980s. Seven models of Westwood garden tractor were made in 1986 with 6 to 16 hp engines and 30 to 42 in. mower decks. The range had increased to twelve lawn tractors by the mid 1990s with 10 to 18 hp

Briggs & Stratton engines.

The hand-propelled Rotary Sickle, introduced in 1959, was the first rotary mower made by John Allen & Sons (Oxford) Ltd, a company famous for the Allen Motor Scythe. The Rotary Sickle, with a four-bladed rotor, was added in 1960. This two-speed, self-propelled mower

Plate 5.42 An early 1970s Mountfield rotary mower with an Aspera engine.

speed 24 in. model and a three speed with a 26 in. rotor were advertised in 1963.

Allen acquired Mayfield Engineering in the mid 1960s and various models of Allen Mayfield mower were manufactured until 1984. Allen Power Equipment continued their own range of domestic rotary mowers including the hand-propelled 18 in. cut Lawnman with a $3\frac{1}{2}$ hp Briggs & Stratton engine, also petrol-engined and mains electric air-cushion mowers.

E.P. Barrus Ltd of Acton in London introduced the Canadian-made two-stroke-engined Lawn-Boy rotary mowers in 1962. Push and self-propelled versions were available in 18 and 21 in. cutting widths with vacuum system to lift the clippings into a cloth grass catcher. The hand-pushed 18 in. model cost £32 10s 0d and the 21 in. Lawn-Boy Automower complete with grass bag was £52.

The Heli-Swift 20 in. rotary grass cutter was sold by Heli-Strand Tools Ltd at Rye in Sussex during the early 1960s. Suitable for lawns, rough grass and scrub, there was a choice of a two- or

cost £95 and advertisements claimed that it made short work of long grass and overgrown vegetation.

The self-propelled $4\frac{1}{2}$ hp Champion rotary mower, launched in 1962, was priced at £140. It was the de luxe model in the Allen range with a three forward speed and reverse gearbox. Models changed quite quickly and three Allen rotary cutters, all with four-stroke engines, comprising a single speed 22 in. machine, a two

Fig 5.14 The first hand-pushed Allen Junior Sickle Mowers were made in 1959 and cost £57 10s 0d.

Fig 5.15   This 1959 Beaver motor scythe had no connection with Beaver Equipment Ltd, bought by Hayter's parent company Tomkins in 1987.

Plate 5.43   Allen Power Equipment Ltd was one of several companies that started making hover mowers when the Flymo patents expired in the mid 1980s.

four-stroke engine for these hand-pushed and self-propelled mowers. An unusual feature of the Heli-Strand 20 was that after removing the drive belts, the engine could be lifted from the mower without using any tools and carried round as a portable power pack for various flexible drive tools including a log saw, hedge trimmer and hand-held rotary tiller.

The Bushwakka motor scythe and scrub cutter, designed to cut really dense growth, was sold by Farmfitters of Gerrards Cross in the late 1950s. It was mounted on large rubber-tyred wheels. An advertisement described the Bushwakka, which

cost £52 10s 0d, as an extremely powerful rotary grass cutter.

Qualcast introduced their first rotary mower, the 18 in. Rotacut with a J.A.P two-stroke engine, in 1957. An improved Mark IV Rotacut with a four-stroke Clinton engine appeared in 1962. The Mark IV, which cost £29 8s 0d, had off-set wheels, a single lever-adjusted cutting height and it was said to use less than one pint of petrol per hour. The Mark V Rotacut with a four-stroke Aspera engine, announced in 1963, was still made in the late 1960s when the price had risen to £33 19s 6d.

Machinery by **Farmfitters Ltd.**

The Versatiller

The Multi-purpose Versatiller brings power-gardening to every home, at a reasonable price. It is the complete handyman. Ideal for private use, also acclaimed by commercial growers. Sturdy, yet exceptionally light to handle.

The Bushwakka

This is a powerful cutter adapted to cut bracken, heather, thorn trees, gorse and similar growth. Sturdily built, yet of light weight. Easy to handle on most difficult ground.

The Rapier Motor

An extremely powerful rotary grass cutter, has an entirely new design. Will tackle without effort the toughest growth. Ideally suitable for the Large Gardens, Parks, Golf Clubs for Field or Lawn.

Gerrards Cross, Bucks

Fig 5.16   Farmfitters included the Bushwakka motor scythe in their early 1960s range of grass-cutting equipment.

Plate 5.44   The 2 hp J.A.P engine on the Qualcast Rotacut, introduced in 1957, used less than one pint of fuel per hour.

The 12 in. cut Rota Mini introduced in 1970 was the first Qualcast mains electric rotary mower. It had a rear roller for mowing over the edge of lawns, a thermal cut-out to protect the motor and a built-in TV interference suppressor. The Rota Mini, complete with 50 ft of cable, cost £11 9s 6d.

The 14 in. cut mains electric Rota Mo 360, added to the Qualcast range in 1971, cost £25 with a grassbox and 72 ft of cable. The Rota Mo, restyled in 1974, was given a rear-mounted grassbox and re-launched as the Qualcast Jetstream. The cheapest 15 in. cut Jetstream with a 1,000 watt mains electric motor and 50 ft of cable or a two-stroke petrol engine cost £50.95. The 18 in. Jetstream with the choice of a $2\frac{1}{2}$ or 3 hp Aspera engine was made for the larger garden and the $3\frac{1}{2}$ hp de luxe version with fold-down handlebars completed the range.

A test report on the electric Jetstream noted that it was impossible for users to get their feet caught under the rotor cover and there was no need to touch the cutting mechanism while it was turning as a red indicator on the motor did not stop until the blades were stationary.

The Suffolk Iron Foundry Polo rotary mower, which cost 25 guineas in the mid 1950s, was replaced by the 18 in. Centaur rotary mower with a four-stroke Clinton Gem and reversible cutter blades in 1959. The Centaur was made until 1963 when it was superseded by the 18 in. Suffolk Galaxy with a two-stroke Aspera engine, off-set wheels and double-sided blades. The colour and cutting performance of the 18 in. Suffolk Meteor announced in 1974 were identical to the Qualcast Jetstream. The standard version of the Meteor with a 3 hp Aspera four-stroke engine cost £58.95 and the $3\frac{1}{2}$ hp Briggs & Stratton-engined de luxe model was £65.95.

Birmid Qualcast were marketing Atco, Qualcast and Suffolk brands by the mid 1970s and three new 18 in. cut petrol-engined Qualcast Suffolk Jetstream rotary mowers were announced in 1979. They replaced the earlier Jetstream and the Atco models, which did not have a grass collector.

Qualcast introduced the Orbital cutting system in 1984. Tough plastic cutters replaced the metal cutter blades that had been used for rotary mowers since the 1930s. A petrol-engined 18 in. Jetstream, also 16 and 19 in. Qualcast Airmo hover mowers, were for larger gardens at the time but the new mains electric Rota-Safe and

Mow-N-Trim rotaries started a new era in the care of small lawns.

The 12 in. cut Rota-Safe lived up to its name with a child-proof safety switch in the handle and the flexible plastic blades reduced the risk of accidentally cutting through the cable. Plastic cutters and nylon line were provided, supplied with the dual purpose Mow-N-Trim; either could be clipped to the rotor for mowing a lawn or trimming around trees, etc.

Suffolk Lawnmowers' mid 1980s rotary mower range included the Rota-Safe, the petrol-engined Jetstream, the Mow-N-Trim, the Atco B45 mulcher mower, also mains electric Qualcast Airmo and petrol-engined Atco Airborne hover mowers. Wolseley Webb, also within the group, were importing a range of Briggs & Stratton-engined Stoic rotary mowers during this period. Stoic mowers were marketed by Landmaster in the 1960s, a 19 in. cut machine with an unusual cutting height adjustment was advertised for

Plate 5.45   The 18 in. Atco rotary mower with a 145 cc four-stroke engine cost £103 11s 0d in 1978.

£47 10s 0d in 1963. Landmaster were acquired by Wolseley Webb in 1980.

The Atco B18 mulcher mower with a 4 hp Briggs & Stratton engine, launched in 1980, cost £175.95. The cut grass was retained under a specially shaped rotor hood while the blades, which made 7,200 cuts per minute, chopped the trimmings to a fine mulch before they were discharged from the rear. The Atco B45 mulcher mower, also with an 18 in. rotor, replaced the B18 in 1984.

Victa (UK) Ltd at Watford imported an Australian-made hand-pushed Victa rotary mower with a 125 cc four-stroke engine and wind-up impulse starter in 1962. It cost £45 and had a two-blade 18 in. cut rotor with free swinging cutters, single lever height adjustment

Plate 5.46   The Landmaster Stoic 19 in. cut rotary mower had a wind-up starter, an unusual wheel-height adjustment and a side-mounted grassbox.

Plate 5.47 The Australian-made Victa VC 160 Auto Drive, introduced in 1974, had a snorkel air filter, dead man's handle to control the driving wheels and a magic eye to indicate when the grassbox was full.

and foldaway handles. These features were not unique but very few rotary mowers made in the early 1960s had all of them. Two paddle blades on the cutting rotor discharged the clippings sideways into an optional side-mounted grass catcher.

An extended range of Victa rotary mowers was imported during the mid 1970s. The 18 in.

cut Chevron with a Victa 125 cc two-stroke engine cost £73 and the Super 24 professional rotary with a Victa 160 cc two-stroke engine was the most expensive and cost £182.

Several companies made small rotary mowers in the early 1960s. The Rotomo made by Barfords of Belton, which cost £30, had a 98 cc two-stroke engine, which was claimed to provide ample power for the 18 in. rotor.

The Boadicea lightweight 14 in. cut rotary mower priced at £23 10s 0d with a tiny 34 cc two-stroke engine was made by H.R. Nash Ltd in Surrey. The handles on the cast aluminium rotor housing could be reversed to cut the sides of ditches and steep banks.

Three models of Scimitar rotary mower were made by Pressure Jet Markers of London in the 1960s. The smallest 8 in. cut mower with a 34 cc two-stroke engine cost £25, the Scimitar 18 and Scimitar Major had 79 cc two-stroke power units and 18 in. cutter heads. Four Seasons Scrub Cutters made at Camberley in Surrey in 18 and 21 in. cutting widths had large rubber-tyred wheels and two-stroke engines. Black & Decker and Wolf Tools were two of the companies who started making small mains electric lawn mowers in the early 1960s. These lightweight mowers sold by hardware stores and garden shops were in direct competition with Qualcast and other specialist grass-care equipment manufacturers.

A British engineer invented the Hovercraft but it was Danish lawn mower manufacturer Karl Dahlman who designed a flying lawn mower without wheels that could float in any direction on a cushion of air. This revolutionary new machine called the Flymo was awarded a gold medal at the 1963 Brussels Inventors' Fair. Great Britain was Europe's largest lawn mower market at the time and for that reason it was decided to manufacture the Flymo hover mower with an Aspera engine at Newton Aycliffe in County Durham. Patents were secured and a small number of hover mowers were made in 1964. Quantity production got under way in 1965 (Colour Plate 23) and the flourishing Flymo business was acquired by Electrolux in 1968. British gardens were getting smaller by the late 1960s and their owners were becoming less energetic when it came to mowing the lawn. This led to a boom in sales of simple electric mowers and Flymo were able to cash in on this trend and the link with Electrolux provided a ready supply of electric motors. There were eight different models of petrol-engined hover mower in the range when the first mains-electric Flymo was introduced in 1969.

Hand-propelled, wheeled 18 in. Flymos with $3\frac{1}{2}$ Briggs & Stratton engines appeared in the mid 1970s together with the Flymo Princess, which did not float. The Princess was a mains electric 15 in. cut cylinder mower with a plastic grassbox

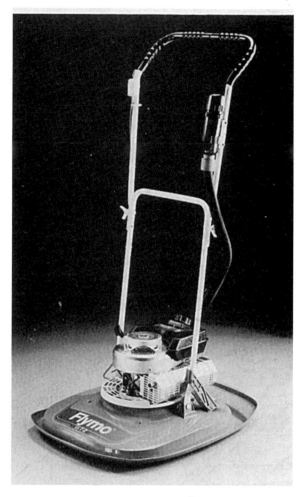

Plate 5.48   Petrol-engined Flymo hover mower.

and roller drive to give the traditional British striped lawn.

The mains electric Flymo DXE, introduced in the late 1970s, was the first hover mower with a grass collector. Ever increasing sales of Flymo hover mowers led to the notorious Mower War in the early 1980s. A Flymo publicity campaign, in its sales drive against the growing popularity of small mains electric cylinder mowers, claimed it to be 'a lot less bover with a hover'. Qualcast countered this in its advertisements for the Concorde mains electric cylinder mower with the slogan 'It's a lot less bover than a hover'.

A year or two later it did not matter any more: the patents protecting the hover principle expired and other companies soon added hover mowers to their product range.

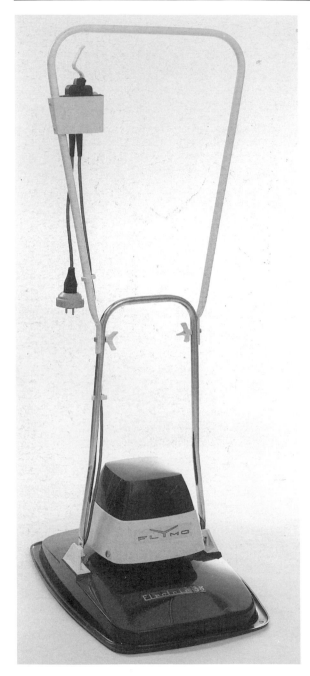

Plate 5.49   The first mains-electric Flymo was made in 1969.

Plate 5.50   The Flymo Turbo Compact 350 was introduced in 1994, thirty years after the first Hover mowers were made in County Durham.

There were eighteen different Flymos in 1981, ranging from a 10 in. mower for under £50 to a 30 in. ride-on model with a floating mower and a seat on wheels for the operator which cost over £800.

Flymo celebrated their 21st birthday in 1985 with the launch of the electric Sprintmaster hover mower. A vacuum system collected the trimmings in a rear grass collector but the smaller clippings were returned to the grass roots to promote growth.

Flymo were also making garden cultivators in the late 1970s and the acquisition of Norlett in 1981 added extra models of wheeled rotary mowers and garden cultivators to their product range.

The Norwegian manufacturer Norlett was selling a range of rotary mowers to British gardeners in the early 1960s. The smallest 19 in. cut hand-pushed model with a $2\frac{1}{4}$ hp Aspera power unit cost £25 10s 0d and the 22 in. self-propelled machine was £65. Norlett made 16 and 19 in. cut rotary mowers with $3\frac{1}{2}$ hp Briggs & Stratton engines for the cheaper end of the market in the mid 1970s. Sales literature described them as tough and dependable with elegant handles plated for protection, which folded for easy storage.

The Flymo Chevron range of wheeled electric

Plate 5.51   This Ariens ride-on mower was one of a range of machines sold by Lely Import, who were appointed UK distributors by the Ariens Company of Wisconsin, USA in 1981.

rotary mowers with a grass collector and rear roller to give the admired striped lawn was launched in 1986. Petrol-engined Chevron mowers were added a year or so later.

Flymo changed their colours from orange and brown to orange and grey and brought the hover principle up to date in 1993 with the Hovervac. Aptly named, the new mower has an integrated grassbox in the body of the mower which is lifted out when it requires emptying.

Ride-on rotary mowers were made for local authority and other professional users but a few small ride-ons for large gardens were introduced in the late 1950s. Many of them, including the 21 in. cut Bolens ride-on, were imported from America. It had a rear-mounted 3 hp engine with rear-wheel drive and handlebar steering on small diameter front wheels.

The Howard Rotavator Co. imported Bolens garden tractors and ride-on mowers in the

Plate 5.52 **The Lawn Flite riding mower with a 5 hp Briggs & Stratton engine was sold with single-speed drive or Auto-drive with speed variable up to 7 mph.**

1970s. The 8 hp Tecumseh-engined Bolens riding mower with a three forward and reverse gearbox and a 28 in. single blade rotary mower was sold by Howard in 1975.

Wolseley Webb ride-on mowers and garden tractors in the late 1970s included a 34 in. cut 348 ride-on mower with an 8 hp Briggs & Stratton engine, four-speed gearbox with a reverse gear and differential. The gold and cream coloured 30 in. Wolseley 308 ride-on was similar but it had automatic drive with a choice of six speeds between 1 and 7 mph. Smaller 26 in. cut machines with 5 and 8 hp engines completed the range.

Allen Power Equipment launched the Motostandard garden tractor in 1961 and continued to distribute the Motostandard/Gutbrod range until 1987. Allen also introduced Roper garden tractors in the UK in 1973. The Allen Roper RT 13 with a 13 hp Briggs & Stratton engine, eight forward and two reverse gears and an

Plate 5.53   Kubota ride-on mower with an 18 hp engine and four-wheel steering.

Plate 5.54   The Allen Motostandard 1010 lawn tractor was made in the early 1970s with an 8 hp Briggs &
Stratton or a 7 hp Tecumseh petrol engine.

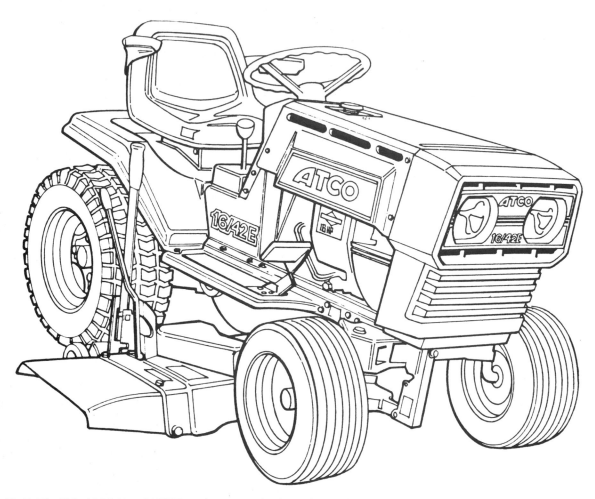

Fig 5.17   This 1984 Atco 16/42E garden tractor had a 16 hp engine and a 32 in. mower deck.

Fig 5.18   The Mountfield 6/25 was typical of early ride-on mowers for the medium-sized garden.

underslung rotary mower deck was typical of American garden tractors imported during the late 1970s.

FINGER BAR MOWERS

Reciprocating knife mowers pulled by horses and tractors have been used on farms for more than 100 years. Smaller machines for garden and smallholdings, either hand-pushed or self-propelled, normally with a 3 ft cutter bar, were introduced in the early 1930s.

The 3 ft cut self-propelled Allen Motor Scythe was made by John Allen & Sons at Oxford for nearly forty years, from 1935 to 1973. The Villiers

Plate 5.55   An advertising slogan for the Allen Motor Scythe declared 'Wherever a man can walk, an Allen will cut'.

for cutting long grass and scrub. Front-mounted attachments included a toolbar, hay sweep, rotary brush, snow plough, yard scraper and cylinder mower. The toolbar was bolted to the cutter bar bracket and could be used to plough, hoe, cultivate and drill seeds. Ancillary equipment for hedge trimming, chain sawing and sheep shearing could be used with either an engine-driven 400 watt generator or a flexible shaft, vee-belt driven from the crankshaft pulley.

Orchard spraying was another job for the Allen Motor Scythe (Plate 4.31) and a trailed seat with a drawbar was used to tow gang mowers, a hay rake and trailer.

Attachments for the 'Plug-In' model F Allen Universal Motor Scythe introduced in 1955 included a 3 ft cutter bar, plough, toolbar, spraying unit and generator. They were plugged into the power unit with two steel arms and the drive shaft for the cutter bar and other powered equipment was connected in a similar way.

Mayfield Engineering at Dorking made three

engine with a cast iron cylinder and piston was rarely any trouble. It sometimes refused to start if there was more than a trace of fuel mixture in the crankcase. This problem was solved by draining it from a drain plug under the engine. A 1.9 hp Villiers two-stroke was used exclusively until 1955 when there was a choice of this engine or a 1.9 hp Villiers Mark 15 four-stroke power unit. A third option was added in 1957 with a 2.9 hp Villiers Mark 25 four-stroke engine. Both four-stroke engines were splash lubricated and the Mark 25 could be converted to run on tvo.

The large-diameter pneumatic-tyred wheels with ratchets in the wheel hubs were driven by a worm and wheel and a pair of reduction gears. The cutter bar was driven from the worm shaft through an enclosed crank to an oscillating spring steel arm located in a socket on the knife back and protected from overload by a safety clutch.

The Allen was much more than a motor scythe

Plate 5.56   The Allen Motor Scythe cutter bar could be offset to achieve a close cut around fruit trees.

Fig 5.19 The Allen 'Plug-in' motor scythe was introduced in the late 1950s.

Plate 5.57   This version of the Allen Mayfield cutter bar mower made in the mid 1970s had an 8 hp Kohler engine.

models of the Mayfield tractor in the late 1950s. They had $1\frac{3}{4}$, $2\frac{1}{2}$ and 3 hp four-stroke engines with a hand lever operated clutch and three-speed gearbox with an optional reverse gear. A front-mounted cutter bar, similar to the Allen Scythe, was one of the fifteen different implements for the tractor unit.

John Allen & Sons acquired Mayfield in the mid 1960s and production of their range of grass-cutting equipment continued under the Allen Mayfield name until 1984.

The Mayfield Eight tractor was used for the Allen Mayfield cutter bar mower made in the 1970s by Allen Power Equipment, formerly John Allen & Sons at Didcot in Berkshire. Power from an 8 hp Kohler engine was used to drive the well-tested Allen cutter bar and propel the mower through a three-speed gearbox with optional reverse to give a top speed of 4 mph.

The cutter bar was driven by a vee-belt, engaged with a jockey pulley clutch system, to a crank and knife arm. A twin-bladed rotary mower or a heavy duty, three-bladed cylinder mower could be fitted in place of the standard 3 ft or optional 4 ft cutter bar in a matter of minutes.

The Allen Mayfield had the same engine in the 1980s but the mechanical gearbox was replaced with a hydrostatic drive unit to give a top speed of 5 mph and 2 mph in reverse. The rotary mower deck was still available but the cylinder mower option had been discontinued.

A work rate eight to twelve times greater than a hand scythe was claimed for the Atcoscythe, introduced in the late 1930s and made for about twenty years. The Atcoscythe, with an output of 2 to 4 acres per day, cost £46 10s 0d in 1939 but the price had risen to £75 by the mid 1950s. The knife and single rubber-tyred wheel were driven by a 150 cc Atco-Villiers two-stroke kick start engine, which used a pint of petrol an hour. Separate levers-operated clutches engaged drive to the cutter bar and land wheel.

An even earlier model, the Atco Autoscythe made during the 1930s, had an auto-cycle type wheel with a Villiers engine to drive the cutter bar.

The Lloyd hand-propelled motor scythe,

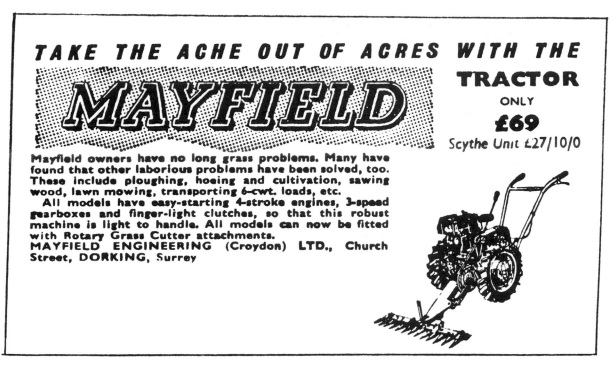

Fig 5.20   The 1962 version of the Mayfield motor scythe.

made in the 1950s, had a single wheel under the engine but later machines had an extra small wheel at the outer end of the cutter bar.

The Jetscythe, made by W.T. Teagle at Truro in the late 1950s and early 1960s, was a single-wheeled self-propelled cutter bar mower with a $\frac{1}{2}$ hp two-stroke Teagle engine and cost £52 10s 0d in 1962.

Cutter bar attachments were made for several two-wheeled garden tractors including the Barford Atom, Clifford, Howard Bantam, Trusty Earthquake, Wolseley Merry Tiller and the American-made Gravely.

A 3 ft cutter bar for the Barford Atom, described in a 1950 catalogue as an invaluable accessory for private gardens, nurseries and estates, cost £17 10s 0d. The cutter bar, also available in 18 in. and 2 ft widths, was centrally mounted on the front of the tractor with an independent clutch to engage the vee-belt drive from the engine crankshaft.

Clifford Aero & Auto produced front- and side-mounted cutter bars for their 1950s range of rotary cultivators. A triple vee-belt drive was used from the gearbox to the front-mounted 3 ft

Plate 5.58   A gallon of fuel was enough to keep the 3 ft cut Atcoscythe running for up to eight hours.

Fig 5.21   W.T. Teagle (Machinery) Ltd made their own engines for the Teagle Jetscythe and the Jetcut hedge trimmer.

or 3 ft 6 in. cutter bar unit on the Model A Mark I cultivator. The knife on the 2 ft 6 in. side-mounted cutter bar for Model A Mark I, Model A Mark III and the Model B rotary cultivators was driven from the rotor shaft by a crank and pitman arrangement. A safety clutch protected the knife from hidden obstructions and the cutter bar was raised with a lever to a vertical position when moving the machine to another site.

A 2 ft off-set front mounted cutter bar was made for Merry Tillers in the 1950s and 1960s. The knife crank was vee-belt driven from the rotary cultivator chaincase with a safety slip clutch to protect it from mechanical damage. A 3 ft cutter bar, called a sickle mower in the Wolseley catalogue, for the more powerful Merry Tillers, which cost £150 in 1977, was made until the very late 1970s.

Cutter bars were available for some garden

Plate 5.59   The Gravely Cutter bar mower attachment for the X-Cel power unit was shown at the 1950 Smithfield Show.

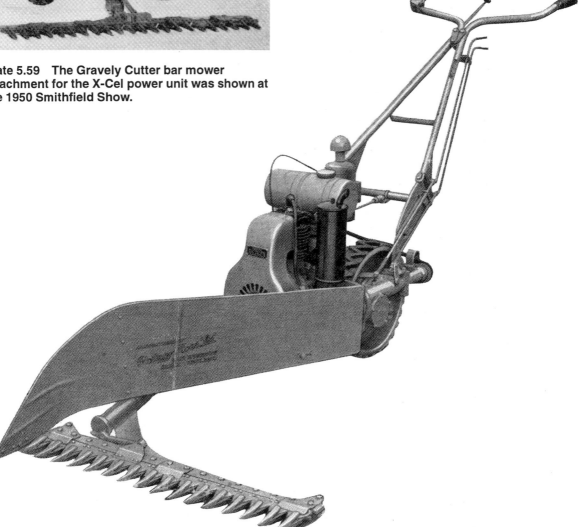

Plate 5.60   The 30 in. cutter bar on this early 1950s Howard Bantam was driven by a countershaft from the rear power take-off used for the cultivating rotor.

Plate 5.61   The Trusty Earthquake with front-mounted cutter bar.

tractors including the Gravely, Howard Dragon and the Iseki cultivator in the early 1980s. Ten years later a front-mounted cutter bar was included in the list of attachments for the Italian-made Bertolini and the Danish Texas garden cultivators.

Gravely Overseas Ltd of Buckfastleigh in Devon introduced the American-built Gravely X-Cel Estate power unit in 1950. It had a $2\frac{1}{2}$ hp four-stroke engine, forward and reverse gears and full differential with an optional diff-lock. It

had a top speed of $2\frac{1}{2}$ mph and a 42 in. front-mounted cutter bar was among the list of power-driven attachments. The early 1950s two-wheeled, 5 hp Gravely Model L garden cultivator could be used with any of twenty-one implements including a knife mower, and a 43 in. cutter bar was made for the Gravely Landworker sold by Gravely Tractors in the early 1960s. Twenty years later Gravely were still offering a 42 in. cutter bar for the 8 and 12 hp 5000 series two-wheel garden tractors.

6 Saws, Trucks and Other Equipment

Circular saws and chain saws were useful tools on the smallholding and market garden and to a certain extent in the garden even in the 1940s and 1950s. Portable saw benches were made at the time for some two-wheeled garden tractors including the Allen Motor Scythe, Farmers' Boy, Merry Tiller and Trusty. The saw was usually on a separate stand and driven by a flat belt from the tractor pulley.

G.W. Wilkin of Kingston-on-Thames exhibited the Farmers' Boy Universal saw bench at the 1950 Smithfield Show. The 12 in. diameter blade was belt driven at 2,000 to 2,500 rpm from a pulley on the tractor power take-off shaft and was capable of cutting through 5 in. thick logs and cut planks up to 4 in. wide. Although it was originally designed for the Farmers' Boy light tractor the saw bench was suitable for any small garden tractor with a power take-off or pulley and it could also be driven with an electric motor.

The Trusty portable saw bench was a more robust item. A chain-driven flat pulley unit attachment was required for the Trusty tractor and after aligning the pulleys on the tractor and saw bench the final task was to secure the bench in position with steel pegs knocked into the ground. Drive to the blade was engaged with a pair of fast and loose pulleys.

The Loyd Motor Sickle and the Treehog were portable circular saws with a difference. Both had a saw blade in front of a two-wheeled frame with an air-cooled petrol engine and handlebars. The Loyd Motor Sickle (Plate 5.31), with a two-stroke Villiers engine, was mainly used for cutting grass and scrub but the makers, Vivian Loyd & Co. of Camberley in Surrey, offered a 14 in. circular saw blade at extra cost. Sales literature claimed that the saw blade could be used to fell tree saplings up to 8 in. diameter and trim protruding tree stumps down to ground level.

Troy Agricultural Utilities, who made a front-mounted rotary scythe for the $1\frac{1}{2}$ hp Troy Tractivator, added an optional 12 in. circular saw blade to the accessory list in 1950. The Tractivator cost £41 10s 0d, the rotary scythe was £9 17s 6d and the saw blade which replaced the scythe disc was an extra £2 17s 6d. Press publicity at the time illustrated the Tractivator with the circular saw in the process of felling a tree but no explanation was offered as to how the operator prevented the tree from falling on to him or the Tractivator!

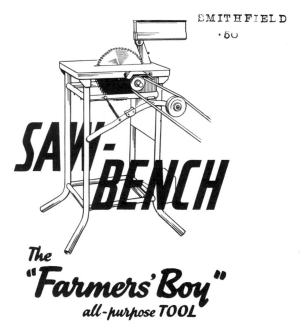

Fig 6.1   The Farmers' Boy saw bench was attached to the tractor for transport purposes.

Plate 6.1   The Trusty tractor power take-off belt pulley could be used to drive various items of stationary equipment including a saw bench, chaff cutter and milking machine.

Plate 6.2   Circular saw attachment for the Loyd Motor Sickle.

Plate 6.3   The Trusty portable saw bench was secured to the ground with steel pegs to make sure it did not move and cause the belt to come off the pulleys.

THE MACHINE WITH A FUTURE

The " Treehog " One-Man Tool

for Tree-felling. Cross-cutting.
Logging and Scrub Clearing.

HOG

A REAL HOG FOR WORK

LANE & WHITTAKER LTD. SHEFFIELD WORKS, HOLYHEAD RD., WELLINGTON, SHROPSHIRE, ENGLAND

Fig 6.2   A Health and Safety inspector would not be impressed by the standard of guarding on the Lane & Whittaker Treehog.

Described as the machine with a future, the Lane & Whittaker Treehog had an adjustable saw head which made it possible to cut down standing trees, cross-cut felled timber and even trim hedges. Two models of Treehog were made in the early 1950s; the Model A with a 3 hp Villiers engine and a 27 in. saw blade cost £110 in 1953. A 5 hp Villiers engine provided the power for the 30 in. diameter blade on the Model B Treehog, which cost £130. The Treehog was wheeled up to a tree with the blade horizontal and according to the manufacturers after the tree was felled, the saw blade could be turned through 90 degrees to make vertical saw cuts and convert the trunk into logs.

CHAIN SAWS

There is some dispute concerning the origin of the chain saw, with Swedish and German companies both staking their claim. It is proven fact that Husqvarna and Stihl were making two-man electric chain saws in the mid 1920s and the first petrol-engined saws appeared in 1927.

The vertical float chamber carburettor on the engine used for early chain saws limited their use to cross-cutting with guide bar always held in a vertical position. Early chain saws had a gearbox to connect the engine to the very long guide bar and required a man at each end of the saw to hold it in position. Automatic chain oiling systems did not appear until 1935 and the cutting chain required frequent attention with an oil can to keep it properly lubricated. Chain saws light enough for one person to use efficiently were not in common use until the mid 1950s. Other milestones in the development of the modern chain saw include an anti-vibration system for the handle in 1965, electronic ignition in 1968, the quick-stop chain brake in 1972 and inertia-operated chain brakes in 1982.

Chain saw buyers in the late 1950s were spoilt for choice with at least twelve makes and numerous models from America, Canada, Germany and Sweden together with British saws made by Danarm, Tarpen and Teles Smith.

Pioneer chain saws with diaphragm carburettors and automatic centrifugal clutches were imported from Canada by E.P. Barrus of London W3. Trojan Ltd of Croydon sold the American-built Clinton saws with 4 and 6 hp engines

Plate 6.4   The Danarm 28B two-man chain saw.

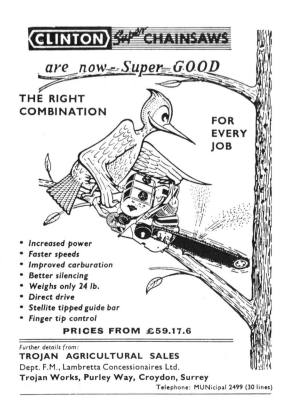

CLINTON *Super* **CHAINSAWS**

are now—Super-GOOD

THE RIGHT COMBINATION

FOR EVERY JOB

- *Increased power*
- *Faster speeds*
- *Improved carburation*
- *Better silencing*
- *Weighs only 24 lb.*
- *Direct drive*
- *Stellite tipped guide bar*
- *Finger tip control*

PRICES FROM £59.17.6

Further details from:
TROJAN AGRICULTURAL SALES
Dept. F.M., Lambretta Concessionaires Ltd.
Trojan Works, Purley Way, Croydon, Surrey
Telephone: MUNicipal 2499 (30 lines)

Fig 6.3   How not to use a Clinton chain saw in 1960—or at any other time!

and 16 to 42 in. guide bars. Other imported chain saws included Dolmar, Homelite, Jo-Bu, Jonsereds, McCulloch, Partner, Remington and Solo.

Homelite saws were made in America and imported by Trojan Ltd in the late 1960s. Homelite, originally concerned with the manufacture of electricity generators, which gave the company its name, made their first one-man chain saw in 1949. Trojan advertised a range of Clinton chain saws with 12 to 42 in. guide bars driven by 58 to 100 cc engines in 1968 and they were still importing these American saws in the mid 1970s.

Clinton chain saws, also made in America, were imported by Trojan Ltd before they became involved with Homelite. Three different saws with 14 to 42 in. guide bars powered by 4 or 6 hp engines were listed in 1960. An 8 hp saw with a similar range of guide bar lengths was added in 1962.

Manufacturers in Canada, Germany and Sweden were the leaders in the development of chain saws in the 1930s but there were others including a British engineer named Armstrong working on their own ideas. Armstrong required production facilities for his ideas so he approached T.H. & J. Daniels, a Stroud engineering company, in 1939 with his drawings for a petrol-engined

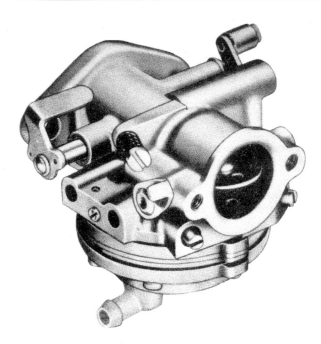

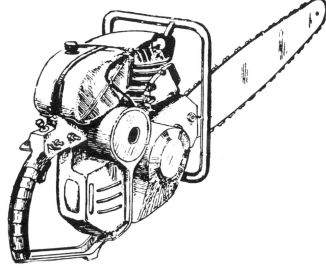

Fig 6.4 The DD 8F was the first Danarm chain saw with a diaphragm carburettor.

Plate 6.5 The Tillotson diaphragm carburettor made it at last possible for direct-drive chain saws to be used at any angle.

chain saw. Daniels and Armstrong agreed to go ahead with the project and the first half of both names were put together to form the trade name Danarm.

The first Danarm chain saw with a 250 cc Villiers two-stroke engine with a vertical float chamber carburettor, made in 1941, was still a two-man saw due to its weight. The earlier problem of not being able to make horizontal cuts was solved with a swivelling mechanism on the gearbox, which allowed a selection of guide bars up to 4 ft long to be rotated. Problems with excessive wear on the guide bar were overcome with hardened steel strips in the chain groove. A larger version, the Danarm 28B with a Villiers 350 cc engine and a choice of guide bars up to 7 ft long, was added at a later date. An improved version of this saw was included in the 1963 Danarm catalogue priced at £145. The 4 ft saw was also built to War Office specifications and some of them were dropped by parachute for use in the dense jungles of Malaya.

Danarm introduced their first one-man saw in 1945. Known as the Danarm Junior Mark I, it had a 90 cc Villiers two-stroke engine and the complete unit was mounted in a tubular metal frame on two wheels. A detachable handle at the outer end of the guide bar allowed two people to use the saw for heavy work. The period between 1946 and 1950 saw the introduction of the Danarm Tornado and Whipper saws. A swivelling rear handle on the Tornado made it possible to use the guide bar at different angles and maintain the float chamber in a vertical position. The Whipper, with an 80 cc J.A.P engine, was a step forward in chain saw design with a recoil starter and automatic chain lubrication.

The Villiers 8F 98 cc two-stroke engine, introduced in 1954, was chosen for the DD8F chain saw launched by Danarm in that year. It weighed about 28 lb and was the Stroud manufacturer's first direct-drive saw. The Villiers engine on the 8F was equipped with a Tillotson diaphragm carburettor, which allowed the engine and guide bar to be used at any angle. The DD8F was made until 1970 but in the meantime the Model 100, with a Danarm 100 cc engine and a choice of guide bar lengths up to 24 in., was added in the early 1960s. Danarm introduced the model 110 saw with their own make 110 cc engine in the early 1960s followed by three 55 cc-engined saws a couple of years later. The Danarm 55, designed to meet a demand for smaller and lighter chain saws, was

Plate 6.6   Danarm DD 8F chain saw.

available with the option of 12, 16 or 21 in. guide
bars. Two standard models with left- or
right-hand recoil starters and a professional saw
with an anti-vibration handle and three shock-
absorbing springs were made for the next
twenty years.

A pneumatic saw with an 18 or 21 in. guide
bar driven by compressed air at a pressure of 80
to 110 psi was in the Danarm range during the
1970s and early 1980s. It weighed about 30 lb
and required a heavy duty compressor and was
therefore of little interest in the horticultural
world but the saw was ideal in dockyards, construction sites, etc. and it could even be used
under water.

An anti-vibration spring suspension system
and a double baffle silencer for low noise levels
were features of the Danarm 1-71-SS saw, with a
71 cc Danarm engine, introduced in 1973. The 12
in. New Frontier saw was for the handyman.
With a 36 cc engine, it weighed 7 lb and cost
about £65. The arrival of the New Frontier
saw coincided with a growing demand for
small, very light saws for the do-it-yourself
enthusiast. Such saws were made in huge numbers mainly in North America; some of these
hobby saws were painted in different colours by
the manufacturer and sold under various names
including Danarm, Pioneer and Dolmar.

Danarm widened their activities in 1979 with
the distribution of the Canadian-made Pioneer
chain saws in the UK. This arrangement continued until 1984 when the production of Danarm saws came to an end and the company was
appointed UK distributor for Zenoah chain saws
from Japan.

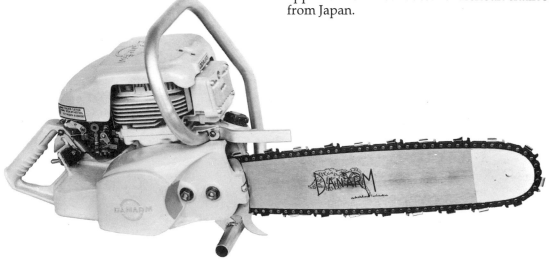

Plate 6.7   A guide bar up to 24 in. long could be used with the 110 cc engined Danarm 110 chain saw.

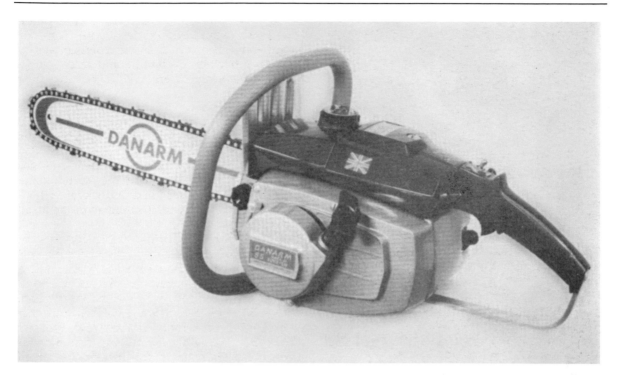

Plate 6.8   The Danarm 55 Mark II left-hand-start chain saw replaced the earlier model in 1972. The 55 was also made with a right-handed recoil starter but it was rather inconvenient when felling trees at ground level.

Plate 6.9   The Danarm engine on the 1-71-SS saw had a 10 to 1 compression ratio and a Tillotson carburettor with a governed speed of 9,500 rpm.

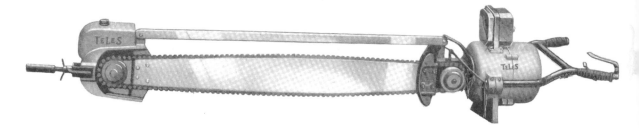

Plate 6.10   The Teles Smith two-man mains electric chain saw.

The company formed by Daniels and Armstrong in 1941 made chain saws for a period of 43 years and for part of that time Danarm was the only British manufacturer of petrol-engined chain saws.

Teles Smith made chain saws at their factory in Undine Street, London SW17 from the 1940s to the early 1970s. A two-man saw with a choice of a 24, 40 or 48 in. guide bar with a 250 cc petrol engine or a 3 hp three-phase electric motor was made by Teles in the 1940s. The Teles Little Tiger one-man saw introduced in the early 1950s was also sold with a petrol engine or electric motor. The petrol-engined saw had an 80 cc J.A.P engine with a float chamber carburettor and the guide bar head could be rotated for felling or logging. There were 240 and 110 volt versions of Little Tiger electric saw; a 15 amp, three-pin socket was suitable for the 240 volt

model but a transformer was needed for the 110 volt saw. It was not always possible to provide a power supply for the electric saws but their great advantage was the ability to use them at any angle.

One- and two-man petrol-engined chain saws with 1 ft to 7 ft guide bars costing from £74 10s 0d to £145 10s 0d were in early 1960 Teles Smith price lists. Teles electric saws with 10 to 30 in. bars cost £44 10s 0d upwards and pneumatic models with 18 and 22 in. guide bars cost £86 10s 0d and £96 10s 0d.

Teles DD77 and the DD95 de luxe direct drive saws with Aspera engines were announced in 1962 and the $5\frac{1}{2}$ hp D95 Super with the option of 18, 22 or 28 in. guide bars was added a year or two later. The D95 Super's Tillotson diaphragm carburettor allowed the saw to be used at any angle and chain drive was automatically engaged

Plate 6.11   The guide bar is locked in the horizontal position while tree felling with the Teles Little Tiger saw.

Plate 6.12   The 28 in. Teles D 95 Super chain saw with a $5\frac{1}{2}$ hp engine weighed 26 lb.

Fig 6.5   The 1967 Stihl model 08S chain saw, with a 5½ hp engine, weighed 14 lb.

by a centrifugal clutch. A chain-saw survey in 1968 concluded that the Teles Wizard with a 1 hp engine and 10 in. bar at £48 was the cheapest on the market and Teles saws like some of their competitors had roller-nosed guide bars. Production of Teles saws ceased in the early 1970s to leave Danarm as the only British chain-saw manufacturer still in business.

A German engineer named Andreas Stihl made a cross-cut saw driven by an electric motor in 1926 and a year later he made a 6 hp petrol-engined tree felling machine which weighed 101 lb. The next step came in 1931 with an 8 hp saw and after the war years, the first Stihl one-man saw, although rather a heavyweight, appeared in 1950. Four years later, Stihl made their first really lightweight chain saw, with a 4½ hp engine, which weighed 31 lb. The centrifugal chain saw clutch, developed by the German company, led to the introduction in

1959 of the Stihl 6 hp gearless drive Contra saw with a diaphragm carburettor.

Stihl chain saws were sold in the UK by Thomas Niven Ltd of Carlisle from the mid 1960s after previously selling the German-made Dolmar saws. The 1968 range of Stihl saws, with guide bars from 13 to 33 in. long, were priced between £71 and £104. Stihl established their own UK marketing organisation in the late 1970s.

There were many makes and models of chain saw available to the professional user by the early 1980s. Since then much improved design has led to lighter and quieter saws with translucent fuel tanks and some even have heated handles. Small electric and petrol-engined saws for the gardener and handyman were improved beyond recognition compared with the lightweight models driven by an electric generator or a flexible driveshaft on a garden cultivator during the 1950s.

Tarpen Engineering Ltd of London NW10 were selling a range of Tarpen-Strand tools in the early 1960s. Designed for flexible drive shafts on garden cultivators they included hedge trimmers, a lawn edger and a 12 in. chain saw. The brand name was changed to Tarpen Flex in the late 1960s and sales literature suggested that the idle power of a motor mower, motor scythe or garden cultivator could be used to drive a range of tools with a Tarpen flexible shaft attachment.

If a suitable engine was not available, Tarpen solved the problem by supplying a flexible shaft

Plate 6.13   A late 1980s Stihl saw.

Plate 6.14   This Tarpen chain saw was driven by a flexible shaft on a lawn mower or garden cultivator engine or attached to a workhead with an electric motor.

Fig 6.6   The Tarpen power take-off attachment for a motor mower.

powered by a Villiers 150 cc four-stroke engine on a small trolley, which could be wheeled around the garden. A work-head power unit consisting of a ¾ hp electric motor with a socket for the flexible shaft was yet another way to drive the chain saw and other tools. The Tarpen chain saw had a 12 in. guide bar and a push-button oiler.

Power for the motor, depending on model, was provided from a 240 volt mains socket, a 110 volt transformer, an engine-driven Tarpen generator or a 12 volt battery in a satchel with a shoulder strap.

HEDGE TRIMMERS

Hooks and shears were the hard way to trim hedges and the reciprocating knife principle in miniature was developed in the late 1940s for trimming garden and field hedges. Attachments for an electric drill or a flexible drive shaft from a garden cultivator or lawn mower, low voltage trimmers powered by a small generator and complete machines with a small engine could be used to reduce the arm-ache caused by a long session with a hook or hedging shears.

Flexible drive shafts for chain saws were equally suitable for hedge trimmers made by Tarpen, Heli-Strand tools and other companies. Electric drill attachments and complete hedge trimmers with an electric drill motor were made by Black & Decker, Stanley-Bridges, Wolf Tools

Plate 6.15   This Black & Decker mains electric chain saw weighed 7½ lb and cost £50.93 when it was introduced in the late 1970s.

and other DIY power tool companies. Engine-driven models such as the Teagle Jet Cut, Teles Clipper from the Teles Division of E.H. Bentall Ltd at Maldon in Essex, the Webb Little Wonder and the more recent Mountfield hedge trimmer had small two-stroke power units.

Webb Little Wonder hedge trimmers were marketed by Wolseley Webb in the 1970s and 1980s. Powered by 110 volt, 240 volt or 12 volt

Plate 6.16   This mid 1960s Tarpen hedge trimmer, with a 12 in. blade, was designed to cut material up to ¾ in. thick.

Plate 6.17   A Hayter 24 in. Motor Scythe with a belt-driven generator and Tarpen hedge trimmer. The generator, which cost £19 15s 0d, could be attached to the scythe in a few minutes.

Plate 6.18   According to a sales leaflet, the electric hedge trimmer for the Allen scythe was ideal for topiary work.

Plate 6.19   Hedge cutting with an electric trimmer powered by a generator on the Trusty tractor.

electric motors or a 21 cc two-stroke petrol engine they had 16 and 30 in. twin reciprocating blades. Mid 1970 prices started at £62 for the 110 volt model with a 16 in. bar and the most expensive 30 in. petrol driven hedge cutter cost £118.

The Allen Motor Scythe was one of many garden mowers and cultivators used with an electric or flexible shaft-driven hedge trimmer in the 1950s. The Allen unit consisted of a 110 volt DC generator with an output of 200 watts, 50 ft of cable and a 12 in. cutter bar driven by a 1/4 hp motor. Sales literature suggested the Allen hedge trimmer could cut hedges as fast as six men using hand shears, also that no particular skill was required and its light weight made it suitable for use by either sex.

The Teagle hedge and weed cutter, first made in 1951, cost £40 delivered to the purchaser's nearest railway station. Drive to the cutter bar was by a vee-belt from a small Teagle two-stroke engine to a secondary roller chain drive running through a tubular frame carried by a strap over the operator's shoulders. The cutter blade could be angled through an arc of 300 degrees for cutting the sides and top of a hedge or weeds at ground level. Sales literature suggested the Teagle hedge cutter would do the work of six men with hooks and although hedge cutting could never be profitable the Teagle machine

made it possible to reduce the cost of this work.

Grass trimmers or strimmers with plastic cord cutters are a relatively recent introduction. Before nylon cord became readily available in the 1970s, mechanical trimming attachments with either a rotary or an endless chain cutter were made by a few companies.

The Stihl brush clearing saw, introduced in the 1950s, was a saw on a stick consisting of a miniature chain saw at one end and a small two-stroke engine near the handles. Ten years later Stihl developed a rotary grass trimmer with a 10 or 12 in. cutting disc driven by a 4 hp engine.

The Tarpen Grassmaster was made in the mid 1960s to cut grass and weeds in places inaccessible to a motor mower or scythe. It consisted of a mains or a battery electric motor on a handle with a set of small rotary cutters for trimming rough grass and lawn edges. Tarpen also made a small rotary cutter head with an 8 in. blade for use with their flexible drive shaft attachment (Figure 6.6) on lawn mowers and garden cultivators.

By the mid 1980s many companies including Allen Power Equipment, Danarm, Homelite, Husqvarna, Mountfield, Sachs Dolmar and Stihl were marketing petrol-engined strimmers of similar design to the Danarm Whipper. Most of

Plate 6.20 The Teagle hedge cutter could be ordered with a two-stroke engine or an electric motor.

Plate 6.22 Hand-held hedge trimmers like this 1992 Kubota 24 cc model are similar to those made twenty years earlier.

Plate 6.21 The Tarpen rotary grass cutter was driven by a flexible shaft from a lawn mower or garden cultivator engine.

them could be used with coarse-toothed metal and tough plastic blades, a nylon cord cutter head and a saw blade. The cutter head was driven by a flexible shaft through the tubular steel handle by a small two-stroke engine.

Small trimmers were made for the domestic gardener in the 1980s by power tool and lawn mower companies. Black & Decker nylon line strimmers replaced hand shears in many gardens during the late 1970s for trimming round trees and posts, up against walls and edges of lawns. Flymo introduced the dual purpose Multi-Trim and the Flymo Mini in 1988. Both machines had a nylon cord cutter driven by a mains electric motor with the cord automatically

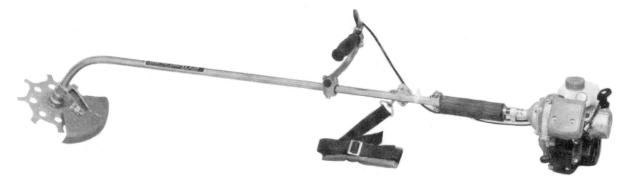

Plate 6.23   The Danarm Whipper lightweight brushcutter was made during the mid 1980s.

fed out to the correct length when the machine was switched off. The cutter head on the Flymo Multi-Trim could be adjusted to trim around trees or edge a lawn.

Qualcast imported the Green Machine from Long Beach, California, in 1988. This was a 1 hp petrol-engined power unit on the handle with a number of attachments including a strimmer, lawn edger, leaf blower and powered weeder with twin reciprocating hoe blades.

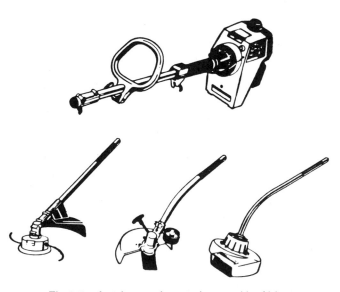

Fig 6.7   A strimmer, lawn edger and leaf blower were among the attachments that could be plugged into the drive shaft from the power unit which formed part of the handle on the Qualcast Green Machine.

MORE GRASS MACHINERY

Lawns require well drained soil with ample plant food to promote growth. Fertiliser was usually spread by hand and a four-tined fork provided a cheap way to improve drainage. Various mechanical aids have been developed over the past fifty years to make these tasks easier.

Sisis Equipment Ltd of Macclesfield was established in 1932 by William Hargreaves, an experienced steam engineer, in some old buildings at Cheadle in Cheshire. The business, which has been mainly concerned with sportsground equipment, moved to larger premises in Macclesfield in 1962. However, some Sisis equipment, including fertiliser spreaders and aerators, has been equally useful on private lawns. Three- and four-pronged steel piercing forks with metal handles, introduced in 1936, were an early Sisis product and cost 16s 6d and 18s 0d respectively. The first Sisis wheeled aerator came a year or two later.

Scarifying the turf to remove thatch and creeping grasses is another important management task for a good lawn and Sisis Rotorake with contra rotating tines was introduced in 1951 for this purpose.

Most Sisis machines were for large areas of sports turf and one in particular caused great interest in the national and international press when it was announced. Although unlikely to have been used even on the largest lawns, the Sisis Drying Machine deserves description. It was taken to Headingley cricket ground in readiness to mop up water from the wicket if rain fell during the 1954 test match against Australia. Newspaper headlines enquired if it

Plate 6.24   The Sisis Rotorake was introduced in 1951.

was really cricket to use the Drying Machine and others awaited the umpires' decision as to whether its use would be legal. A leaflet published by W. Hargreaves & Co. Ltd of Sisis Works, Cheadle quoted part of a paragraph from the Laws of Cricket which stated that any roller may be used for drying the wicket and making it fit for play.

The Drying Machine consisted of a steel drum covered with an easily removable absorbent material and adjustable rollers which squeezed the water into tanks at the front and rear of the machine. The drum could be suitably weighted with two spring-loaded pistons to suit conditions and the water tanks were easily removed for emptying. The machine was transported on two pneumatic-tyred wheels, which could be retracted when it was in use.

Several County cricket clubs and the All England Lawn Tennis Club used the Sisis Drying machine in 1953 and Essex CCC were said to have taken theirs on tour with the team.

E. Allman & Co. of Chichester, noted for their spraying equipment, introduced a rather different-looking lawn roller in 1954. Although the Feasa looked like a roller it was a fertiliser and lawn sand spreader, which was said to lay an even carpet of any dry material on a lawn. There were two models, one for the home gardener and a larger machine for bowling greens and sports grounds. The smaller Feasa spreader cost £6 5s 6d, it weighed 72 lb when filled with sand and could be adjusted to apply from 1 to 8 oz of sand per sq yd when pushed at a speed of 2 mph. The sports ground model weighed 182 lb with a full load and cost £12 15s 0d.

Simple hand-operated lawn spikers including the hollow-tined Sheen lawn aerator were pushed into the turf like a fork. A more expensive version, which cost £9.90 in the mid 1970s, had spring-loaded hollow tines and as they were pushed into the lawn plugs of soil passed upwards inside the tines to be collected in a detachable tray.

Plate 6.25   The Sisis drying machine for cricket wickets caused quite a stir on the sports pages of newspapers during the summer of 1954.

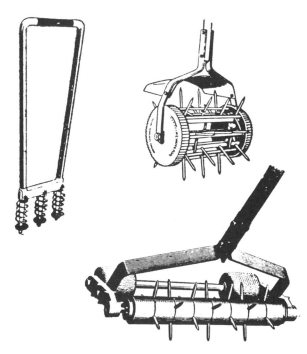

Fig 6.8   Lawn aerators. The spring-loaded tines on the late 1950s Sheen lawn aerator removed cores and left them on the surface. The Jalo roller aerator made in the early 1960s cost £4 15s 0d and the Tudor spiker, which made 150 holes per minute, cost £7.50 in 1973.

An advertisement in 1962 for the Jalo roller aerator suggested that it made easy work of aerating a lawn with steel spikes making 50 holes per second when pushed over the turf. Another type of roller aerator with a number of tine bars between two wheels was pushed across the lawn to make holes about $1\frac{1}{2}$ in deep. The Tudor aerator was an example of this design, the 9 in. wide model cost £6.75 in 1973 and the 12 in. machine was £7.50.

The autumn chore of sweeping up leaves is another task that has attracted the attention of the inventive mind. Hand-propelled sweepers made by Allen, Atco, Wolf and other companies, with a brush to throw the leaves into a suitable container hooked on the handlebars, have been available for many years. Much wider leaf sweepers, notably those made by Allen in the 1970s, were self-propelled or towed by a small tractor. A $2\frac{1}{2}$ hp Briggs & Stratton engine was used to propel the 30 in. wide, pedestrian controlled Allen Scorpio sweeper, which cost £120 in the mid 1970s.

A rotary brush provided another method of sweeping leaves into a swath for collection and this could also be achieved with a leafblower. The Ransomes Leafblower with a fan driven by an 8 hp Briggs & Stratton engine was one such

Plate 6.26   The Iseki garden tractor and rotary brush made short work of sweeping paths.

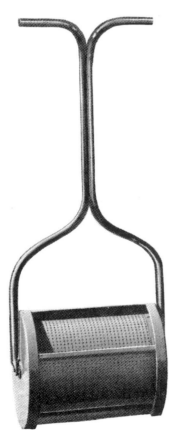

Plate 6.28   The leaf catcher on the Atco garden sweeper could be tipped to empty the contents into a wheelbarrow.

machine. Sales literature for the mid 1980s Ransomes machine indicated that the $\frac{1}{4}$ in. thick fan blades would clear leaves from walkways and leave them in rows for collection.

Atco came up with an alternative method of tidying the garden with the introduction of the

Plate 6.27   The Allman Feasa spreader was used to spread fertiliser, sand, etc. on lawns.

Plate 6.29   Leaves were blown into heaps for collection with the Ransomes leaf blower.

Plate 6.30   This low-loading trailer was made by Whitlock Brothers of Great Yeldham in Essex for the Ransomes MG crawler.

Blow & Vac in 1987. A 31 cc engine and fan unit was used to suck leaves and debris into a large capacity bag carried on the shoulder or to blow leaves into convenient piles for hand collection. The Blow & Vac was discontinued in 1989.

Flymo launched the Garden Vac, described as a portable vacuum cleaner, in 1993. A 650 watt mains electric motor drives the fan, which can be set to blow leaves into piles and then switched to the vacuum position to suck them into a bag on the machine. The vacuum mode has two settings for light and heavy garden debris.

TRAILERS AND TRUCKS

Manure, compost, soil, garden rubbish and probably firewood was and still is usually moved around a garden in a wheelbarrow.

However smallholders and market gardeners, who were not inclined to push a barrow, could invest in a motorised truck or a trailer for their garden cultivator. Several types of motor truck including the Batric Truc Tractor, Geest Tug, Martin, Trac Grip, Tubo Truc Tracta and the Wrigley were used by gardeners, nurserymen and smallholders during the 1950s. These useful little vehicles served their purpose for many years and similar machines, such as the tracked Kubota transporter, are made in the mid 1990s.

The lightweight three-wheeled Wrigley motor truck with a Villiers engine on a turntable above the single front wheel was invented by Arthur Wrigley, assisted by his brother, in the early 1930s. Arthur Wrigley founded Wessex Industries at Poole in Dorset a year or so later to manufacture the truck. Production restarted after the war years and by 1949 two models of Wrigley truck with a 1 hp 98 cc Villiers Mark 10 and a 3 hp 250 cc four-stroke side valve petrol engine with a variety of different bodies were made. The 1 hp truck with an Albion three-speed gearbox and chain drive to the front wheel, which cost about £95, was able to carry 10 cwt on level ground but the load capacity was reduced in less favourable situations. The tubular steel handlebar was hinged so that the user could either ride on the truck or lead it from the front. The rear wheels had rod linkage brakes operated with a foot pedal, the hand brake was on the front wheel and the truck bed on the tipper version was raised hydraulically. The 3 hp truck had a Burman gearbox with three forward speeds and reverse, a top road speed of 10 mph and a maximum load of 1 ton.

The Wessex Industries product range included fork trucks, trailers and wheelbarrows when the

Fig 6.9   A water cart saved many steps when watering a large number of plants. This model was advertised as a new product in 1954.

Fig 6.10   The 1974 version of the Wrigley Motor Truck was made by Wantage Engineering Ltd.

company ceased trading in the late 1960s. The business was purchased by fork lift manufacturers Montgomerie Reid of Basingstoke, who eventually sold the motor truck side to Wantage Engineering in 1974. The Wrigley Union name was revived and production of ride-on and pedestrian-controlled ½ and 1 ton trucks began at Wantage. Factories, warehouses and horticultural enterprises were the main customers for the truck, which was hardly changed from the original design, but there was a choice of body styles and petrol, diesel or electric power units.

Either a Honda petrol engine or a 7 hp air-cooled Lister diesel provided power through a clutch to a three forward and reverse gearbox on Wrigley trucks built in the early 1980s. They were still steered in the same way but an upholstered driving seat provided far more comfort than in earlier days when the driver sat on a short wooden plank or a suitable wooden box. Wantage Engineering still manufacture a

much improved version of the three-wheeled Wrigley Truck but still in the same style as those made by Arthur Wrigley sixty years earlier.

The first Truc-Tractas were made in 1951 by Tubo Metal Products at Pontypridd in Glamorgan and like the Wrigley truck, the user could either lead it or ride on the seat. It was advertised in 1951 at £79 ex works and extremely economical carting was claimed to be one of its advantages. Power to haul a load of up to 8 cwt was provided by a 1¼ hp J.A.P four-stroke engine through a two-speed gearbox to the front wheel. The steering handle also engaged the clutch and front-wheel brake.

A wider range of Truc-Tractas was made in 1954 with prices starting at £35 for the Junior model. The standard machine was £65 and a 10 cwt capacity Estate truck cost £89 10s 0d. The J.A.P engine was replaced with a 2 hp B.S.A. engine at this time, and the standard Truc-Tracta had an Albion gearbox with two forward gears

Fig 6.11   The brake was applied to the front wheel when the driver released the steering handle on the Tubo Truc-Tracta, made in 1957.

Plate 6.31   The Bonser Truck cost £216 in the early 1950s. The bumper bar around the front wheel was an extra £4 5s 0d.

and reverse with roller chain drive to the front wheel and had a top speed of 8 mph. In common with the 8 cwt model, the steering handle on the Estate version operated the clutch but the expanding shoe brakes were pedal operated and a hand brake was provided for parking. The Junior Truc-Tracta with a $1\frac{1}{4}$ hp two-stroke J.A.P 80 engine was said to be the cheapest motor truck on the market at the time. The 4 ft by 2 ft 6 in. load platform could carry up to 5 cwt at a maximum speed of 3 mph and a tipping platform was available at extra cost.

Hatch Brothers of Callington in Cornwall made three-wheeled Trac Grip motor trucks in the 1950s. Improved versions were introduced from time to time including a 10 to 12 cwt capacity model in 1954. Unlike some of the earlier trucks, drive was from a 2 hp or optional $3\frac{1}{2}$ hp B.S.A. engine through a differential rear axle to the rear wheels. The 1954 Trac Grip tipper truck, which cost £170, had a cushioned seat for the driver and unlike most other trucks at the time it had a steering wheel instead of the more usual handlebars, with a chain and sprocket linking it to the front wheel.

The 1950s Geest Tug was, as its name suggests,

a towing unit rather than a truck but it did have a small load-carrying platform behind the driving seat. A Villiers air-cooled engine and an Albion three forward and reverse gearbox were mounted on a turntable above the single front wheel.

Fig 6.12   The rear wheels could be removed to convert the Nash Roller-Tractor onto a 4 cwt roller. The consolidating effect was increased by placing weights in the dumper hopper.

Plate 6.32   The John Deere AMT (all-materials transporter) in the motor truck of the 1990s.

The Batric Tugtractor made at Stroud in the early 1970s was a three-wheeled electric motor truck that could travel 6 miles with a fully charged 12 volt battery. The range could be extended by adding one or two extra batteries to give a range of either 12 or 18 miles before recharging was necessary. The Tugtractor was not a tug, it resembled a handcart with the motor and batteries beneath the load platform. Early models had a 12 cu. ft capacity dropside or tipper body and an improved model with a lower chassis was introduced in 1974. The Tugtractor, on pneumatic tyres with a forward and reverse control unit, cost about £200. An infinitely variable electronic speed control system was available at extra cost. The truck was steered by a pedestrian, with a tubular handle linked to the single rear castor wheel and by using either forward or reverse gear the

operator could either walk in front of or behind the truck. A socket outlet on the Tugtractor provided a power source for auxiliary lighting and power tools including a battery-operated hedge trimmer. Attachments included a front-mounted scraper blade and a detachable tipping bin for carrying fertilisers, manure, etc.

More powerful trucks including the Bonser and Opperman Motocart carried much heavier loads. The 5 hp Bonser three-wheeled tipper truck made by Bonser Equipment Ltd of Hucknall near Nottingham had a 13 ft 6 in. turning circle and when fully laden it carried one ton. The specification included three forward speeds and reverse, rear-wheel drive through a hypoid differential axle and 9 in. Girling shoe brakes.

The Martin 1 ton truck with a hand or hydraulically tipped body with a 7 or 10 hp Kohler engine was made in the mid 1970s by

Malcolm Martin Trucks of Kirby in Ashfield in Nottinghamshire. A shortened version of the 12 hp tricycle-wheeled motor truck with a rear drawbar was used to tow a trailer and other equipment.

The 30 cwt Opperman Motocart, made in the 1950s, had an air-cooled 8 hp engine at one side of its large diameter front driving wheel. The whole unit was mounted on a turntable and provided the means of steering the Motocart, which had a top speed of 12 mph.

Although intended for building work, small dumper trucks were used by some market gardeners to move goods and materials around the holding. The 1950s Nash Roller-Tractor was a dumper with a difference in that with the rear wheels it could be used as a land or road roller. Power from a 420 cc B.S.A. G Series air-cooled petrol engine was transmitted by roller chain to three forward and reverse Albion gearbox and a second roller chain linked the gearbox output shaft to the rear wheels. Driving controls consisted of a lever-operated clutch with a pedal to disengage drive to the off-side rear wheel, the front wheel was steered with a handlebar and the foot brake was locked on for parking.

The hover truck principle was used by the

Plate 6.33   The Allen Motor Scythe with a load carrier.

Light Hovercraft company of East Grinstead during the mid 1970s for the Hover-Pallet, said to be able to traverse almost any terrain on a cushion of air. It had a flat-load platform for sacks, bales, etc. and depending on model could carry up to 10 cwt. The Light Hovercraft Co. also made a small self-propelled tracked load carrier at the time. The Catterbug with a 2 cwt payload was driven by a 171 cc four-stroke engine with

Plate 6.34   This four-wheeled trailer was made for the Clifford Mark IV cultivator.

Plate 6.35   Reinforced sides, corners and floor were features of this steel trailer for Trusty garden tractors.

a twist grip throttle and an automatic clutch. Drive was by vee-belt direct from the engine pulley to a large diameter pulley on the drive shaft for the wide slatted track.

The pedestrian-controlled Kubota tracked carrier (Colour Plate 39) is a very sophisticated machine compared with the Catterbug twenty years earlier. It has a 5.5 hp Kubota petrol engine, two forward and reverse gears and can carry about 6 cwt.

Speed and comfort were added attractions of motorised trucks by the mid 1980s. The John Deere AMT with high flotation tyres, four-wheel drive and steered by a fifth wheel at the front will go virtually anywhere. It has a variable speed belt driven transmission, optional hydraulically tipped body and seats for driver and passenger.

Two-wheeled garden tractors of all sizes were used to move materials around in a trailer in the 1940s and 1950s. A 7½ cwt trailer with internal expanding shoe brakes was made by Barford (Agricultural) Ltd for the little Barford Atom 15 garden tractor (Plate 1.23). A wooden platform seat was provided for the driver and the catalogue pointed out that the side and end boards could be removed to facilitate low loading. A steel-sided 5 cwt capacity tipping dump truck with its own parking brake was also made for the Barford Atom.

A carrier attachment made for the Allen Motor Scythe during the 1950s also served as a platform for chemical tank when spraying fruit trees with a hand lance (Plate 4.31). The load platform was secured above the wheels with four bolts and a small wheel, mounted on the cutter-bar bracket, supported the front end. With front board removed the carrier could be tipped to empty out bulk materials.

Plate 6.36   The Merry Truck cost £25 when it was added to the list of Merry Tiller accessories in 1960.

Plate 6.37   The Trusty dumper truck shows a family resemblance to the Trusty Steed tractor. It had a 5 hp J.A.P engine and a $\frac{1}{3}$ cu yd skip.

Plate 6.38   Massey Ferguson 1010 garden tractor and trailer.

Plate 6.39   A Ford 1920 compact tractor and trailer transporting bales of peat around a plant nursery.

Wolseley Engineering added the Merry Truck to a long list of equipment for the Merry Tiller in 1960. Most garden cultivator trucks were towed behind the tractor with a seat for the person steering the outfit. The Merry Truck was different as it had a circular frame attached to the Merry Tiller and secured with a single pin. Once in position, the Merry Tiller could be turned through a full circle within the frame so that it could either pull or push the truck. The operator walked behind to steer the Merry Truck with the cultivator handlebars when it was pushed along and could walk alongside or ride in the trailer when it was towed. The Merry Truck could carry up to 5 cwt and the body could be tipped to empty out bulk loads.

Tractors (London) Ltd, well known for Trusty two- and four-wheeled tractors, widened their product range in 1957 with the addition of contractor's plant including a roller, angle dozer and dumper trucks. The dumpers, with many of the parts the same as those used for Trusty Steed

tractors, had skip capacities of $\frac{1}{3}$ and 1 cu. yard. There was a choice of a 6 hp air-cooled single cylinder J.A.P or a Petter AVA 1 diesel engine for both models, which had a three forward and reverse gearbox, independent wheel clutches for sharp turns and a drawbar for towing other equipment.

Trailers for small garden tractors held little more than a barrow load but they were made bigger to keep in step with the increasing size of the more powerful three- and four-wheeled ride-on tractors such as the Gunsmith, Garner and Trusty Steed. This trend continued with the advent of compact tractors and even larger trailers were made for them. However, as so often is the case, the need for small tractors and trailers for large gardens was ignored in the rush to make bigger and better equipment. Eventually a new generation of garden tractors and ride-on mowers equipped with a rear towbar arrived in the 1970s and trailers were made to match.

Index

223

FARMING PRESS BOOKS

Below is a sample of the wide range of agricultural and veterinary books published by Farming Press. For more information or for a free illustrated book list please contact:

Farming Press Books
Wharfedale Road, Ipswich IP1 1RJ, United Kingdom
Telephone (01473) 241122   Fax (01473) 240501

Fifty Years of Farm Machinery
BRIAN BELL
A highly illustrated, detailed account of the history of British farm machinery from the 1940s to today.

Farm Machinery
BRIAN BELL
Gives a sound introduction to a wide range of tractors and farm equipment. Now revised, enlarged and incorporating over 150 photographs.

Farm Workshop
BRIAN BELL
Describes the requirements of the farm workshop and illustrates the uses of the necessary tools and equipment.

Machinery for Horticulture
BRIAN BELL & STEWART COUSINS
A description of the basic functions and uses of the diverse machinery in all aspects of horticulture.

Ford and Fordson Tractors
Massey-Ferguson Tractors
MICHAEL WILLIAMS
Heavily illustrated guides to the models which made two leading companies great.

New Hedges for the Countryside
MURRAY MACLEAN
Gives full details of hedge establishment, cultivation and maintenance for wind protection, boundaries, livestock containment and landscape appearance.

Farm Welding
ANDREW PEARCE
A highly illustrated guide to stick welding, gas welding and cutting, MIG/MAG techniques, soldering and basic blacksmithing.

Tractors Since 1889
MICHAEL WILLIAMS
An overview of the main developments in farm tractors from their stationary steam engine origins to the potential for satellite navigation.

Farming Press Books is part of the Morgan-Grampian Farming Press Group which publishes a range of farming magazines: *Arable Farming, Dairy Farmer, Farming News, Pig Farming, What's New in Farming.* For a specimen copy of any of these please contact the address above.